# Mathematics for Technicians New Level II

**A. Greer** CEng, MRAeS
FORMERLY SENIOR LECTURER

**G. W. Taylor** BSc (Eng), CEng, MIMechE
PRINCIPAL LECTURER

Gloucestershire College
of Arts and Technology

STANLEY THORNES (PUBLISHERS) LTD

First published in 1982 by:
Stanley Thornes (Publishers) Ltd
Old Station Drive
Leckhampton
CHELTENHAM GL53 0DN
England

Reprinted with minor corrections 1984
Reprinted with minor corrections 1985
Reprinted 1986


British Library Cataloguing in Publication Data

Greer, A.
    Mathematics for technicians Level II.—New ed.
    1. Shop mathematics
    I. Title      II. Taylor, G.W.
    510'.246      TJ1165

ISBN 0-85950-353-4

Typeset by
Tech-Set, Gateshead, Tyne and Wear
Printed and bound in Great Britain at
The Bath Press, Avon

# CONTENTS

# AUTHORS' NOTE

This volume covers the TEC Level II half-units:

| | |
|---|---|
| U80/691 | Mathematics |
| U80/692 | Analytical Mathematics |
| U80/712 | Mensuration |

Also included are some basic topics, such as the use of electronic calculating machines, which should prove extremely useful for students (with exemption from Level I Mathematics) who enter Level II directly.

The format adopted will prove ideal for colleges who decide to base their Level II Mathematics on topics taken from all three of the standard half-units mentioned above.

A Greer
G W Taylor                                                    Gloucester

# THE SCIENTIFIC ELECTRONIC CALCULATOR

*After reaching the end of this chapter you should be able to:*

1. Understand accuracy in calculations due to rounding and truncation.
2. Understand the need for a rough check answer.
3. Obtain a rough check answer for any calculation.
4. Use the calculator to find the values of arithmetic expressions involving addition, subtraction, multiplication, division and use the memory.
5. Extend operations to include reciprocals, square roots and numbers in standard form.
6. Evaluate arithmetical expressions involving whole number, fractional and negative indices.
7. Use the calculator to evaluate formulae.
8. Evaluate polynomials using nested multiplication.
9. Use the calculator to find the values of trigonometrical functions.

## ACCURACY IN CALCULATIONS

Two phrases in common use are 'decimal places' and 'significant figures'.

*Decimal places*: These refer to the number of figures after the decimal point, e.g. 35.1 has one decimal place, and 0.0035 has four decimal places.

*Significant figures*: These are the number of figures counted from the left to the right, starting with the first *non-zero* figure. The following numbers have three significant figures: 407, 20.3, 1.67, 0.662, 0.0190, 0.005 16

## ROUNDED NUMBERS

The process of 'rounding' or 'rounding-off' enables a degree of accuracy to be stated. Thus rounding 3167 gives 3170 correct to three significant figures, and rounding 3163 gives 3160 correct to three significant figures.

### Rule for Rounding

If the figure to be discarded is less than five then the previous figure is *not* altered, but if the figure to be discarded is greater than five then the previous figure is increased by one.

1

Thus 54 276 may be stated as 54 280 correct to four significant figures or 54 000 correct to two significant figures.

Also 3.174 92 may be stated as 3.175 correct to four significant figures or 3.17 correct to three significant figures.

## ACCURACY OF NUMBERS

The number 52 is considered to be correct to two significant figures which means it has been rounded from between 51.5 and 52.5. This may be stated as $52 \pm 0.5$ showing that the maximum error is $\pm 0.5$

Similarly 6.2 is considered accurate to two significant figures and lies between $6.2 \pm 0.05$. Also 0.1370 is considered accurate to four significant figures and lies between $0.1370 \pm 0.000\,05$

In many problems, such as calculations on a triangle, the lengths of the sides may be given as 6, 8 and 9 metres. Here, three figure accuracy should be assumed and the calculations performed assuming the dimensions are 6.00, 8.00 and 9.00 metres

## ACCURACY IN ADDITION AND SUBTRACTION

Consider the problem $67 - 62$ which, assuming two significant figure accuracy, may be stated as $(67 \pm 0.5) - (62 \pm 0.5)$

Now $67 \pm 0.5$ has a minimum value of 66.5 and a maximum value of 67.5. Also $62 \pm 0.5$ has a minimum value of 61.5 and a maximum value of 62.5

Thus the least difference is $66.5 - 62.5 = 4.0$
and the greatest difference is $67.5 - 61.5 = 6.0$

Hence the result lies between 4 and 6 and may be stated as $5 \pm 1$

> In general when adding and subtracting numbers the maximum error of the result may be found by adding the maximum errors of the original numbers.

Thus $620 + 56.3$ implies $(620 \pm 0.5) + (56.3 \pm 0.05)$

$$= (620 + 56.3) \pm (0.5 + 0.05) = 676.3 \pm 0.55$$

# ACCURACY IN MULTIPLICATION AND DIVISION

Consider the problem $67 \times 62$ which, assuming two figure accuracy as previously, will give

the least value of the product $= 66.5 \times 61.5 = 4089.75$

and the greatest value of the product $= 67.5 \times 62.5 = 4218.75$.

Thus the result of $67 \times 62$, when rounded to two significant figures, will lie between 4100 and 4200 and cannot therefore be guaranteed to two figure accuracy.

However it is generally accepted (despite not being strictly correct) that if:

When multiplying and dividing numbers the answer should *not* be given to an accuracy greater than the least accurate of the given numbers.

Thus the calculated result of $67 \times 62$, which is 4154, should be stated as 4200 which has been rounded to two significant figures.

Consider also $\dfrac{573 \times 21}{6240}$ the result of which is $1.928\,365\,3$ on a calculator. The least accurate of the given numbers is 21 which has two significant figure accuracy. Hence the answer must not be stated to more than two figure accuracy, namely 1.9.

# TRUNCATION OR CUTTING-OFF

Most calculating machines 'truncate' or 'cut off' numbers in their displays after computation. For instance, in an eight figure display the result of 5 divided by 3 is shown as $1.666\,666\,6$ which contains a recurring figure 6 and if the result had been rounded would be $1.666\,666\,7$. Now truncation occurs each time a computation is performed and this may introduce an accumulating error.

# ROUGH CHECKS

When using a calculator it is essential for you to do a rough check in order to obtain an approximate result. Any error, however small it may seem, in carrying out a sequence of operations will result in a wrong answer.

Suppose, for instance, that you had £1000 in the bank and then withdrew £97.82. The bank staff then used a calculator to find how much money you had left in your account—they calculated that £1000 less £978.2 left only £21.80 credited to you. You would be extremely annoyed and probably point out to them that a rough check of £1000 less £100 would leave £900, and that if this had been done much embarrassment would have been avoided.

The 'small' mistake was to get the decimal point in the wrong place when recording the money withdrawn, which is typical of errors we all make from time to time. You should get in the habit of doing a rough check on any calculation *before* using your machine. The advantage of a rough check answer before the actual calculation avoids the possibility of forgetting it in the excitement of obtaining a machine result. Also your rough check will not be influenced by the result obtained on your calculator.

## KEYBOARD LAYOUT

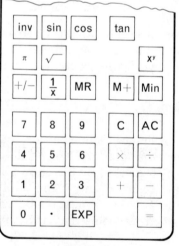

Fig. 1.1

The following are the figure keys:

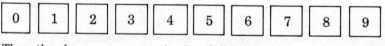

The other keys are summarised as follows:

| . | decimal point key. |

| EXP | EXPONENTIAL KEY—allows entry of numbers in standard form (e.g. $1.46 \times 10^3$). |

| C | clear key—enables an incorrect entry to be erased (either a figure or an operation) if pressed immediately afterwards. |

| AC | all clear key—clears machine of all numbers—except the memory. |

| $\times$ | $\div$ | $+$ | $-$ | $=$ |    arithmetical operation keys.

| Min | memory in key—enters a number on display into the memory, erasing any number previously in the memory. |

| M+ | add to memory key—enables a number to be added to previous content of memory. |

| MR | memory recall key—enables content of memory to be shown on display. |

| +/− | change sign key—(e.g. $+2$ to $-2$, or $-1.5$ to $+1.5$). |

| $\frac{1}{x}$ | reciprocal key (e.g. $\frac{1}{2}$ to 2, or 4 to 0.25). |

| $\pi$ | 'pi' key—gives the numerical value of $\pi$. |

| $\sqrt{}$ | square root key. |

| $x^y$ | power key—gives the values of $y$ to the index $x$. |

| sin | cos | tan |    trigonometric function keys—gives the sine, cosine or tangent of an angle.

| inv | inverse trigonometric function key—gives the angle when given the value of a trigonometrical ratio, e.g. $\text{inv}\sin\theta$ (also called $\arcsin\theta$ or $\sin^{-1}\theta$). |

Part of a typical keyboard layout on an electronic calculator is shown in Fig. 1.1. Calculators vary in layout and operation, just as motor cars do from different manufacturers, but the methods of using each type are similar. Each calculator is supplied with an instruction booklet and you should work through this carrying out any worked examples which are given.

In this chapter we have outline procedures which are generally common to all calculators—if they are not exactly as your machine requires, you will have to make allowance according to the instruction booklet.

## WORKED EXAMPLES

After first switching on the calculator, or commencing a fresh problem, you should press the $\boxed{\text{AC}}$ key. This ensures that all figures entered previously have been erased and will not interfere with new data to be entered.

The memory is not cleared but this is done automatically when a new number is entered in the memory using the $\boxed{\text{Min}}$ key.

**EXAMPLE 1.1**

Evaluate  $18.24 + 4.39 - 9.72$

A rough check gives:    $18 + 5 - 10 = 13$

The sequence used on the calculator is similar to the order in which the problem is given. This is shown by:

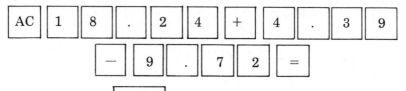

The display gives $\boxed{12.91}$ and since all the data is given to two decimal places the answer, correct to two decimal places is $12.91$

It should be noted that the order of operations is not important— try for yourself the sequence $18.24 - 9.72 + 4.39$ and you will obtain the same answer.

**EXAMPLE 1.2**

Evaluate $\dfrac{20.3 \times 3092}{1.563}$

A rough check gives: $\dfrac{20 \times 3000}{2} = 30\,000$

The sequence of operations is:

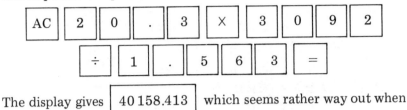

The display gives [ 40 158.413 ] which seems rather way out when compared with the approximate answer 30 000 obtained from the rough check. This is not unexpected, however, since we did allow for 2 in the denominator instead of 1.563 which would make the rough check answer smaller. It does confirm that the true answer is of the correct order.

It would not be correct to state the answer as 40 158.413 given on the display, since the least accurate of the given numbers, 20.3, has three significant figures and this is the accuracy to which we should state the answer.

Thus the answer is 40 200 correct to three significant figures.

### EXAMPLE 1.3

Find the value of $6\,857\,000 \times 119\,000 \times 85.3$

For the rough check numbers which contain as many figures as these are better considered in standard form:

$$(6.857 \times 10^6) \times (1.19 \times 10^5) \times (8.53 \times 10)$$

or approximately:

$$(7 \times 10^6) \times (1 \times 10^5) \times (10 \times 10) = 7 \times 1 \times 10 \times 10^{12}$$
$$= 70 \times 10^{12}$$
$$= 7 \times 10^{13}$$

The sequence of operations is:

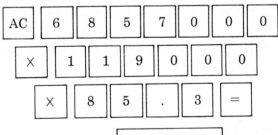

The display will show [ 6.960 33    13 ] which represents $6.960\,33 \times 10^{13}$. The least number of significant figures in the given numbers is three. Thus the answer is $6.96 \times 10^{13}$.

An alternative method is to enter the numbers in standard form using the EXP (exponential) key. The sequence would then be:

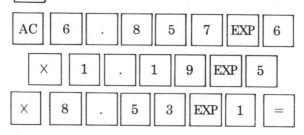

giving the same result.

The sequence used in a problem such as this would be personal choice, but if the problem includes numbers with powers of 10 the latter sequence is better.

### EXAMPLE 1.4

Evaluate $\dfrac{(5.745 \times 10^3)(56.7 \times 10^{-4})}{0.0343 \times 10^6}$

The rough check gives: $\dfrac{(6 \times 10^3)(6 \times 10^{-3})}{3 \times 10^4} = \dfrac{6 \times 6}{3} \times 10^{3-3-4}$

$$= 12 \times 10^{-4}$$

The following sequence includes use of the +/− (change sign) key to enter the negative index:

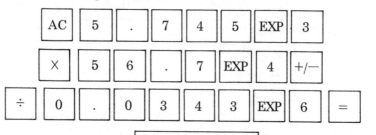

The display will show $\boxed{9.496\,84 \qquad -04}$ and since the least number of significant figures in the given numbers is three, then the answer is $9.50 \times 10^{-4}$.

Note that we give the answer stating 9.50 and not merely 9.5 which would imply only two significant figure accuracy.

**EXAMPLE 1.5**

Evaluate $\dfrac{0.674}{1.239} - \dfrac{0.564 \times 1.89}{0.379}$

The rough check will need a little more care when approximating numbers—it is not possible to give rules but as you gain experience you will have no difficulty.

Thus the rough check gives  $\dfrac{0.5}{1} - \dfrac{0.5 \times 2}{0.5} = 0.5 - 2 = -1.5$

The sequence of operations includes use of the memory in which intermediate results may be kept for use later in the sequence. You should note also how a number may be subtracted from the contents of the memory by using the $\boxed{M+}$ (add to memory) key after changing the sign of the number.

The sequence of operations is:

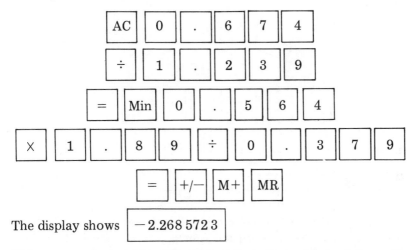

The display shows  $\boxed{-2.268\,572\,3}$

This answer is considerably higher than that obtained from the rough check, but it is of the correct order, i.e. *not* 22.7 *or* 0.227

Hence the required answer is $-2.27$ correct to three significant figures.

**EXAMPLE 1.6**

Find the value of:  $9.7 + \dfrac{55.15}{29.6 - 8.64}$

The rough check gives: $10 + \dfrac{60}{30-9} \approx 10 + \dfrac{60}{20} = 13$

It is possible to work this problem out on the calculator by re-arranging and using the memory. However no rearrangement is necessary if we make use of the $\boxed{\dfrac{1}{x}}$ (the reciprocal) key.

This key enables us to find the reciprocal of a number—for example, the reciprocal of 2 is $\frac{1}{2}$ or 0.5

Let us consider $9.7 + \dfrac{1}{\left(\dfrac{29.6-8.64}{55.15}\right)}$

This may be written as:

$$9.7 + 1 \div \left(\dfrac{29.6-8.64}{55.15}\right) = 9.7 + 1 \times \left(\dfrac{55.15}{29.6-8.64}\right) = \dfrac{55.15}{29.6-8.64}$$

If we make use of this knowledge then the sequence of operations is:

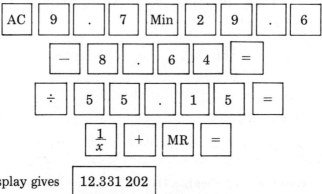

The display gives $\boxed{12.331\ 202}$

It is always difficult to assess accuracy of the answer to a calculation which involves addition and subtraction. Although the 9.7 has only two significant figures, the addition of the other portion of the calculation will increase the figures before the decimal point to two.

Thus the answer may be given as 12.3

In most engineering problems you will not often be wrong if you give answers to three significant figures—this is consistent with the accuracy of much of the data (such as ultimate tensile strengths of materials). There are exceptions, of course, such as certain machine shop problems which may require a much greater degree of accuracy.

## 'WHOLE NUMBER' POWERS

The obvious method, but not necessarily the quickest, to find the fifth power of 5.6 (for example) is to multiply 5.6 by itself four times. To avoid entering the number five times use may be made of the memory as the following sequence shows:

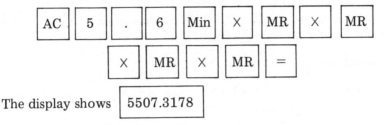

The display shows | 5507.3178 |

Thus $(5.6)^5 = 5510$ correct to three significant figures.

## USE OF THE 'CONSTANT MULTIPLIER' FACILITY

You should check with the calculator booklet that this is possible with your machine.

Suppose that we have the following to evaluate:

$3.1 \times 7.89$,  $3.1 \times 6.2$,  $3.1 \times 3.45$,  $3.1 \times 9.8$,  $3.1 \times 10.9$

If each calculation is carried out individually, then the number 3.1 will have to be entered for each calculation. This may be avoided by two consecutive pressings of the | × | key. Thus the sequence would be:

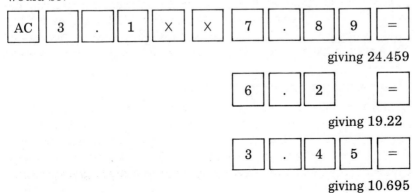

giving 24.459

giving 19.22

giving 10.695

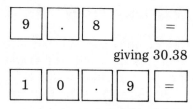

giving 30.38

giving 33.79

If in the above sequence, entry of each of the numbers preceding the equals operation were omitted, then the calculator would automatically substitute 3.1 for each of them.

Thus $(3.1)^4 = 3.1 \times 3.1 \times 3.1 \times 3.1$ may be found by the sequence:

| AC | 3 | . | 1 | × | × | = | = | = |

giving 92.3521

## Exercise 1.1

Evaluate the following:

1) $45.6 + 3.563 - 21.42 - 14.6$

2) $-23.94 - 6.93 + 1.92 + 17.6$

3) $\dfrac{40.72 \times 3.86}{5.73}$

4) $\dfrac{4.86 \times 0.008\,34 \times 0.64}{0.86 \times 0.934 \times 21.7}$

5) $\dfrac{57.3 + 64.29 + 3.17}{64.2}$

6) $\dfrac{32.2}{6.45 + 7.29 - 21.3}$

7) $\dfrac{1}{\frac{1}{3} + \frac{1}{4} + \frac{1}{5}}$

8) $\dfrac{3.76 + 42.4}{1.6 + 0.86}$

9) $\dfrac{4.82 + 7.93}{-0.73 \times 6.92}$

10) $9.38(4.86 + 7.6 \times 1.89^3)$

11) $4.93^2 - 6.86^2$

12) $(4.93 + 6.86)(4.93 - 6.86)$

13) $\dfrac{1}{6.3^2 + 9.6^2}$

14) $\dfrac{3.864^2 + 9.62}{3.74 - 8.62^2}$

15) $\dfrac{9.5}{(6.4 \times 3.2) - (6.7 \times 0.9)}$

16) $1 - \dfrac{5}{3.6 + 7.49}$

17) $\frac{1}{6} - \frac{1}{5}(4.6)^2$

18) $\dfrac{6.4}{20.2}\left(3.94^2 - \dfrac{5.7 + 4.9}{6.7 - 3.2}\right)$

19) $\dfrac{3.64^3 + 5.6^2 - (1/0.085)}{9.76 + 3.4 - 2.9}$

20) $\dfrac{6.54(7.69 \times 10^{-5})}{0.643^2 - 79.3(3.21 \times 10^{-4})}$

# SQUARE ROOT AND 'PI' KEYS

 gives the square root of any number in the display.

$\boxed{\pi}$ gives the numerical value of $\pi$ to whatever accuracy the machine is designed.

**EXAMPLE 1.7**

The period, $T$ seconds (the time for a complete swing), of a simple pendulum is given by the formula $T = 2\pi\sqrt{\dfrac{l}{g}}$ where $l$ m is its length and $g$ m/s$^2$ is the acceleration due to gravity.

Find the value of $T$ if $l = 1.37$ m and $g = 9.81$ m/s$^2$.

Substituting the given values into the formula we have

$$T = 2\pi\sqrt{\frac{1.37}{9.81}}$$

The rough check gives: $T = 2 \times 3\sqrt{\frac{1}{9}} = 2 \times 3 \times \frac{1}{3} = 2$ s

The sequence of operations would be:

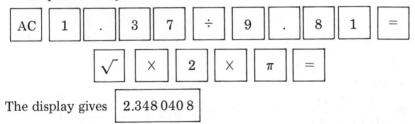

The display gives $\boxed{2.348\,040\,8}$

Thus the value of $T$ is 2.35 seconds, correct to three significant figures.

# THE POWER KEY

$\boxed{x^y}$ gives the value of $x$ to the index $y$.

**EXAMPLE 1.8**

The relationship between the luminosity, $I$, of a metal filament lamp and the voltage, $V$, is given by the equation $I = aV^4$ where $a$ is a constant. Find the value of $I$ if $a = 9 \times 10^{-7}$ and $V = 60$

Substituting the given values into the equation we have

$$I = (9 \times 10^{-7})60^4$$

The rough check gives

$$I = (10 \times 10^{-7})(6 \times 10)^4 = 10^{-6} \times 6^4 \times 10^4 = 10^{-2} \times 36 \times 36$$

and if we approximate by putting $30 \times 40$ instead of $36 \times 36$

then     $I = 10^{-2} \times 30 \times 40 = 10^{-2} \times 1200 = 12$

The sequence of operations would be:

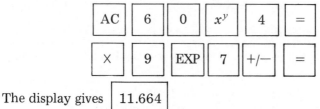

The display gives   11.664

Thus the value of $I$ is 11.7 correct to three significant figures.

### EXAMPLE 1.9

The law of expansion of a gas is given by the expression $pV^{1.2} = k$ where $p$ is the pressure, $V$ is the volume, and $k$ is a constant. Find the value of $k$ if $p = 0.8 \times 10^6$ and $V = 0.2$

Substituting the given values into the formula we have

$$k = (0.8 \times 10^6) \times 0.2^{1.2}$$

The rough check gives

$$k = 1 \times 10^6 \times (\tfrac{2}{10})^{1.2} = 10^6 \times \frac{2^{1.2}}{10^{1.2}} = 10^6 \times \frac{3}{30} = 0.1 \times 10^6$$

Since it is difficult to assess the approximate value of a decimal number to an index, it becomes simpler to express the decimal number as a fraction using whole numbers. In this case it is convenient to express 0.2 as $\tfrac{2}{10}$. We can guess the rough value of $2^{1.2}$, since we know that $2^1 = 2$ and $2^2 = 4$. Similarly we judge the value of $10^{1.2}$ as being between $10^1 = 10$ and $10^2 = 100$. The more practice you have in doing calculations of this type, the more accurate your guess will be.

The sequence of operations would be:

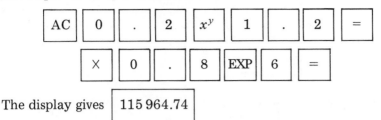

The display gives    115 964.74

Thus the required value of $k$ is 116 000 or $0.116 \times 10^6$ correct to three significant figures.

### EXAMPLE 1.10

The intensity of radiation, $R$, from certain radioactive materials at a particular time, $t$, follows the law $R = 95t^{-1.8}$. If $t = 5$ find the value of $R$.

Substituting $t = 5$ into the given equation gives $R = 95 \times 5^{-1.8}$

For a rough check on the value of an expression containing a negative index it helps to rearrange so that the index becomes positive.

The rough check gives: $R = 100 \times 5^{-2} = 100 \times \dfrac{1}{5^2} = 100 \times \dfrac{1}{25} = 4$

The sequence of operations would be:

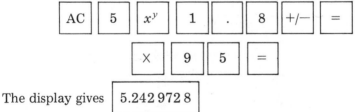

The display gives    5.242 972 8

Hence the value of $R$ is 5.24 correct to three significant figures.

### EXAMPLE 1.11

In the formula $Q = 2.37H^{5/2}$ find $Q$ when $H = 2.81$

Substituting $H = 2.81$ into the given formula $Q = 2.37(2.81)^{5/2}$

The fractional index should be expressed as a decimal so that we may use the $x^y$ key. Since $\frac{5}{2} = 2.5$ then equation may be stated as $Q = 2.37(2.81)^{2.5}$

The rough check gives   $Q = 2 \times 3^{2.5} = 2 \times 15 = 30$

We estimated the value of $3^{2.5}$ as 15 since it lies between $3^2 = 9$ and $3^3 = 27$

The sequence of operations is:

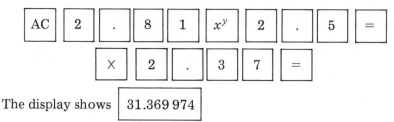

The display shows   $\boxed{31.369\,974}$

Hence  $Q = 31.4$  correct to three significant figures.


# NESTING (OR STACKING) A POLYNOMIAL

A polynomial in $x$ is an expression containing a sum of terms, each term being a power of $x$.

A typical polynomial is

$$ax^4 + bx^3 + cx^2 + dx + e$$

where  $a, b, c, d$ and $e$  are constants.

The polynomial may be factorised successively as follows:

$$ax^4 + bx^3 + cx^2 + dx + e$$
$$= (ax + b)x^2 + cx^2 + dx + e$$
$$= \{(ax + b) + c\}x^2 + dx + e$$
$$= [\{(ax + b) + cx\} + d]x + e$$

The opening bracket symbols $[\{($ are usually omitted and all the remaining closure brackets are written in the same form. Thus the polynomial looks like:

$$ax + b) + c)x + d)x + e$$

This is called the nested (or stacked) form of the polynomial. When evaluating its value we must always work from *left* to *right*.

**EXAMPLE 1.12**

Find the value of $y = 5x^3 - 7x^2 + 8x - 5$ when $x = 3.32$

The nested form gives $y = 5x - 7)x + 8)x - 5$

When $x = 3.32$ then $y = 5 \times 3.32 - 7)3.32 + 8)3.32 - 5$

Working from *left* to *right* the sequence of operations is:

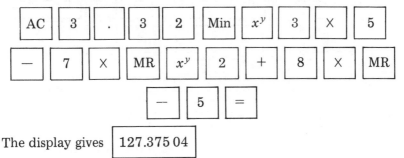

The display gives $\boxed{127.375\ 04}$

Thus $y = 127$ correct to three significant figures.

Note that if the polynomial is not nested it would be necessary to evaluate

$$5(3.32)^3 - 7(3.32)^2 + 8(3.32) - 5$$

Try this for yourself and you will appreciate the advantage of nesting.

# TRIGONOMETRICAL FUNCTIONS

Use of $\boxed{\sin}$, $\boxed{\cos}$, $\boxed{\tan}$ and $\boxed{\text{inv}}$ keys.

**EXAMPLE 1.13**

Find angle $A$ if $\sin A = \dfrac{3.68 \sin 42°}{5.26}$

*Rough check*: It is always difficult to find an approximate answer for a calculation involving trigonometrical functions. However, we may use a 'backwards substitution' method which is carried out after an answer has been obtained.

On most calculators there is a sliding switch which may be positioned at either RAD (radians), DEG (degrees) or GRAD (grades—a grade being one-hundredth of a right angle, used more on the continent). In this example the angles are in degrees and so we set the sliding switch to the DEG position.

A sequence of operations commencing with

| AC | 3 | . | 6 | 8 | X | 4 | 2 | sin | = |

may not function correctly as it merely gives the value of sin 42°. Try it on your own machine. We must, therefore, alter the order in which the operations are carried out. Thus the sequence used should be:

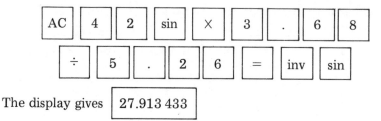

| AC | 4 | 2 | sin | X | 3 | . | 6 | 8 |

| ÷ | 5 | . | 2 | 6 | = | inv | sin |

The display gives $\boxed{27.913\ 433}$

Thus angle $A = 27.9°$ correct to three significant figures.

*Answer check.* As the result of our calculations we have

$$\sin 27.9° = \frac{3.68 \sin 42°}{5.26}$$

Now we may rearrange this expression to give

$$\sin 42° = \frac{5.26 \sin 27.9°}{3.68}$$

Thus if we find the value of the right-hand side of this new expression, it should give the value of sin 42° if the value of angle $A = 27.9°$ is correct.

The sequence of operations is similar to that given for the original calculation. Try it for yourself and thus check the result.

## Exercise 1.2

Evaluate the expression in Questions 1–16 giving the answers correct to three significant figures.

1) $2.32^4$      2) $1.52^6$     3) $0.523^5$     4) $7.9^{-2}$

5) $4.59^{-3}$     6) $0.321^{-4}$     7) $12.1^{1.5}$     8) $6.83^{2.32}$

9) $0.879^{3.1}$     10) $5.56^{0.62}$     11) $14.7^{0.347}$     12) $3.9^{-0.5}$

13) $6.64^{3/2}$     14) $13.6^{2/5}$     15) $1.23^{7/3}$     16) $0.334^{3/5}$

17) Evaluate $4\pi r^2$ when $r = 6.1$

18) Evaluate $5\pi(R^2 - r^2)$ when $R = 1.32$ and $r = 1.24$

19) In a beam the stress, $\sigma$, due to bending is given by the expression $\sigma = \dfrac{My}{I}$. Find $\sigma$ if $M = 12 \times 10^6$, $y = 60$, and $I = 11.5 \times 10^6$

20) The polar second moment of area, $J$, of a hollow shaft is given by the equation $J = \dfrac{\pi}{32}(D^4 - d^4)$. Find $J$ if $D = 220$ and $d = 140$

21) The velocity, $v$, of a body performing simple harmonic motion is given by the expression $v = \omega\sqrt{A^2 - x^2}$. Find $v$ if $\omega = 20.9$, $A = 0.060$, and $x = 0.02$

22) The natural frequency of oscillation, $f$, of a mass, $m$, supported by a spring of stiffness, $\lambda$, is given by the formula

$f = \dfrac{1}{2\pi}\sqrt{\dfrac{\lambda}{m}}$. Find $f$ if $\lambda = 5000$ and $m = 1.5$

23) The volume rate of flow, $\dot{Q}$, of water through a venturi meter is given by

$$\dot{Q} = A_2 \sqrt{\frac{2gH}{1 - \left(\dfrac{A_2}{A_1}\right)^2}}$$

Find $\dot{Q}$ if $A_1 = 0.0201$, $A_2 = 0.005\,03$, $g = 9.81$, and $H = 0.554$

24) A non-dimensional constant used in connection with rectangular weirs is given by $\dfrac{H^{3/2}g^{1/2}}{n}$. Find the value of this constant when $H = 4.56$, $g = 32.2$ and $n = 8.42 \times 10^{-6}$

25) The specific speed of a centrifugal pump is given by the expression $\dfrac{N\sqrt{Q}}{H^{3/4}}$. Find the specific speed when $N = 2000$, $Q = 140$ and $H = 7$

26) The specific speed of a hydraulic turbine is given by $\dfrac{N\sqrt{P}}{H^{5/4}}$. Find the specific speed if $N = 150$, $P = 30 \times 10^3$ and $H = 20$

27) Find the value of $a$ if $a = \dfrac{80.6 \sin 55°}{\sin 70°}$

28) If $\cos C = \dfrac{a^2 + b^2 - c^2}{2ab}$, find the value of the angle $C$ when $a = 19.37\,\text{mm}$, $b = 26.42\,\text{mm}$ and $c = 22.31\,\text{mm}$

29) The following formula is used when measuring the angle of a vee notch: $\sin \dfrac{\theta}{2} = \dfrac{\frac{1}{2}(D-d)}{H - h - \frac{1}{2}(D-d)}$. Find $\theta$ when $D = 38.10\,\text{mm}$, $d = 25.40\,\text{mm}$, $H = 100.92\,\text{mm}$ and $h = 78.03\,\text{mm}$

30) When checking the major diameter of a metric thread the following formula is used: $M = D - \dfrac{5p}{\tan 30°} + d\left(1 + \dfrac{1}{\sin 30°}\right)$. Find $M$ when $p = 5.30\,\text{mm}$, $D = 18.34\,\text{mm}$ and $d = 15.38\,\text{mm}$

31) Use the method of nesting to evaluate the following:
(a) $5x^2 + 4x - 15$ when $x = 3.8$
(b) $7x^3 + 3x^2 - 7x + 5$ when $x = 0.35$
(c) $x^4 + 3x^3 - 8x^2 + x - 3$ when $x = 3.75$
(d) $x^3 + 5x^2 - 6x + 9$ when $x = 1.2$

# 2. TABLES OF VALUES

---

After reaching the end of this chapter you should be able to:

1. Draw up a suitable table of values as an orderly approach to carrying out repeated calculations from an equation or formula.

---

## TABLES OF VALUES

When calculating a series of values for an expression it is essential to have an orderly approach and a concise method of recording results. A table of values may be used for this purpose—the following examples show typical applications.

### EXAMPLE 2.1

If $y = x^3 + 2x^2 + 3x - 7$ find the values of $y$ for $x = -3$ to $x = +3$ at intervals of 1 unit.

The following table of values enables each term to be evaluated separately for each value of $x$. The corresponding values of $y$ are then found by adding up the relvant terms in each column.

| $x$ | $-3$ | $-2$ | $-1$ | $0$ | $1$ | $2$ | $3$ |
|---|---|---|---|---|---|---|---|
| $x^3$ | $-27$ | $-8$ | $-1$ | $0$ | $1$ | $8$ | $27$ |
| $2x^2$ | $18$ | $8$ | $2$ | $0$ | $2$ | $8$ | $18$ |
| $3x$ | $-9$ | $-6$ | $-3$ | $0$ | $3$ | $6$ | $9$ |
| $-7$ | $-7$ | $-7$ | $-7$ | $-7$ | $-7$ | $-7$ | $-7$ |
| $y$ | $-25$ | $-13$ | $-9$ | $-7$ | $-1$ | $15$ | $47$ |

### EXAMPLE 2.2

Find the values of $z$ for values of $\theta$ from $0°$ to $75°$ at $15°$ intervals if $z = 2 \sin \theta + 3 \cos \theta$.

In this table you should note the 'staggering' of the columns which enables the two relevant terms, i.e. $2\sin\theta$ and $3\cos\theta$ to be added conveniently without the interference of the values of $\sin\theta$ and $\cos\theta$.

| $\theta°$ | 0 | 15 | 30 |
|---|---|---|---|
| $\sin\theta$ | 0 | 0.259 | 0.500 |
| $2\sin\theta$ | 0 | 0.518 | 1.000 |
| $\cos\theta$ | 1 | 0.966 | 0.866 |
| $3\cos\theta$ | 3 | 2.898 | 2.598 |
| $z$ | 3 | 3.416 | 3.598 |

| $\theta°$ | 45 | 60 | 75 |
|---|---|---|---|
| $\sin\theta$ | 0.707 | 0.866 | 0.966 |
| $2\sin\theta$ | 1.414 | 1.732 | 1.932 |
| $\cos\theta$ | 0.707 | 0.500 | 0.259 |
| $3\cos\theta$ | 2.121 | 1.500 | 0.777 |
| $z$ | 3.535 | 3.232 | 2.709 |

## EXAMPLE 2.3

Find the values of $5\sin(2\phi + 72)°$ for $\phi = 0°$, $10°$, $20°$, $30°$, $40°$, and $50°$.

Again the calculations may be tabulated conveniently as shown — line four values are obtained by adding lines two and three together.

| $\phi°$ | 0 | 10 | 20 | 30 | 40 | 50 |
|---|---|---|---|---|---|---|
| $2\phi°$ | 0 | 20 | 40 | 60 | 80 | 100 |
| $72°$ | 72 | 72 | 72 | 72 | 72 | 72 |
| $(2\phi + 72)°$ | 72 | 92 | 112 | 132 | 152 | 172 |
| $\sin(2\phi + 72)°$ | 0.951 | 0.999 | 0.927 | 0.743 | 0.469 | 0.139 |
| $5\sin(2\phi + 72)°$ | 4.755 | 4.995 | 4.635 | 3.715 | 2.347 | 0.695 |

**EXAMPLE 2.4**

Find the overall average speed of a car over four stages of a motor rally if:

stage 1 of 10 km distance was completed at a steady speed of 50 km/h,

stage 2 of 30 km distance was completed at a steady speed of 20 km/h,

stage 3 of 20 km distance was completed at a steady speed of 35 km/h, and

stage 4 of 15 km distance was completed at a steady speed of 40 km/h.

We know that

$$\text{Average speed} = \frac{\text{Distance travelled}}{\text{Time taken}}$$

and rearranging

$$\text{Time taken} = \frac{\text{Distance travelled}}{\text{Average speed}}$$

We will now tabulate the given data and find the time taken for each stage.

| *Stage* | *Stage length* (km) | *Steady Speed* (km/h) | *Stage time* $= \dfrac{Stage\ length}{Steady\ speed}$ (h) |
|---|---|---|---|
| 1 | 10 | 50 | $\dfrac{10}{50} = 0.2$ |
| 2 | 30 | 20 | $\dfrac{30}{20} = 1.5$ |
| 3 | 20 | 35 | $\dfrac{20}{35} = 0.571$ |
| 4 | 15 | 40 | $\dfrac{15}{40} = 0.375$ |
| | 75 Total distance | | 2.646 Total time |

Thus:    Overall average speed $= \dfrac{\text{Total distance travelled}}{\text{Total time taken}}$

$$= \frac{75}{2.646} = 28.3 \text{ km/h}$$

# Exercise 2.1

Answer the following questions by drawing up suitable tables of values.

1) If $y = 6x^2 - 3x - 9$ find the values of $y$ for $x = -2$ to $+2$ at intervals of 0.5

2) Find the values of the expression $6 \tan (3\phi + 35)^\circ$ for $\phi = 0^\circ$ to $\phi = 30^\circ$ at intervals of $5^\circ$.

3) Find the values of $V$ if $V = 5 \cos \theta - 7 \tan \theta$ for values of $\theta$ equal to $0^\circ, 10^\circ, 20^\circ, 30^\circ, 40^\circ, 50^\circ$ and $60^\circ$.

4) The acceleration, $a$, of the piston of an engine is related to the angle $\theta$, which the crank makes with the centre-line of the cylinder by the expression $a = 98.6 (\cos \theta + 0.25 \cos 2\theta)$. Find the values of $a$ if $\theta$ has values from $0^\circ$ to $140^\circ$ at $20^\circ$ intervals.

5) On a shopping spree five shirts cost £6.50 each, seven cardigans cost £32.00 each, three skirts cost £27.60 each, and six pairs of jeans cost £22.65 each. Find the average cost of a garment.

# TRANSPOSITION OF FORMULAE

## TRANSPOSITION OF FORMULAE

In the formula $P = I^2R$, $P$ is called the subject of the formula. It may be that we are given values of $I$ and $P$ and we have to find $R$. We can do this by substituting the given values in the formula and solving the resulting equation.

We may have several sets of corresponding values of $I$ and $P$ and want to find the corresponding values of $R$. Much time and effort will be saved if we express the formula with $R$ as the subject because then we need only substitute the given values of $I$ and $P$ in the rearranged formula.

The process of rearranging a formula so that one of the other symbols becomes the subject is called *transposing the formula*. The rules used in the transposition of formulae are the same as those used in solving equations. The methods used are as follows:

## Symbols Connected as a Product

EXAMPLE 3.1

Transpose $W = IV$ to make $V$ the subject.

Divide both sides by $I$:

$$\frac{W}{I} = \frac{IV}{I}$$

or

$$\frac{W}{I} = V$$

∴

$$V = \frac{W}{I}$$

25

**EXAMPLE 3.2**

Make $R$ the subject of the formula $H = I^2RT$.

Divide both sides by $I^2T$:

$$\frac{H}{I^2T} = \frac{I^2RT}{I^2T}$$

or

$$\frac{H}{I^2T} = R$$

$\therefore$

$$R = \frac{H}{I^2T}$$

# Symbols Connected as a Quotient

**EXAMPLE 3.3**

Transpose the formula $R = \dfrac{V}{I}$ to make $V$ the subject.

Multiply both sides by $I$:

$$R \times I = \frac{V}{I} \times I$$

or

$$RI = V$$

$\therefore$

$$V = RI$$

**EXAMPLE 3.4**

Transpose the formula $R = \dfrac{\rho l}{a}$ to make $a$ the subject.

Multiply both sides by $a$:

$$R \times a = \frac{\rho l}{a} \times a$$

$\therefore$

$$Ra = \rho l$$

Divide both sides by $R$:

$$\frac{Ra}{R} = \frac{\rho l}{R}$$

$\therefore$

$$a = \frac{\rho l}{R}$$

# Symbols Connected by a Plus or Minus Sign

**EXAMPLE 3.5**

Transpose $T = t + 273$ for $t$.

Subtract 273 from both sides:

$$T - 273 = t$$

or
$$t = T - 273$$

**EXAMPLE 3.6**

Transpose $V = E + IR$ for $I$.

Subtract $E$ from both sides:

$$V - E = IR$$

Divide both sides by $R$:

$$\frac{V - E}{R} = \frac{IR}{R}$$

or
$$\frac{V - E}{R} = I$$

$\therefore$
$$I = \frac{V - E}{R}$$

# Formulae Containing Brackets

**EXAMPLE 3.7**

Transpose $I = \dfrac{V}{R + x}$ for $V$.

Multiply both sides by $(R + x)$:

$$I \times (R + x) = \frac{V}{R + x} \times (R + x)$$

or
$$I(R + x) = V$$

$\therefore$
$$V = I(R + x)$$

**EXAMPLE 3.8**

Transpose $R = R_0(1 + \alpha t)$ for $t$.

Remove the bracket on the RHS:   $R = R_0 + R_0 \alpha t$.

Subtract $R_0$ from both sides:   $R - R_0 = R_0 \alpha t$.

Dividing both sides by $R_0 \alpha$:

$$\frac{R - R_0}{R_0 \alpha} = \frac{R_0 \alpha t}{R_0 \alpha}$$

$$\therefore \qquad t = \frac{R - R_0}{R_0 \alpha}$$

**EXAMPLE 3.9**

Transpose $x = \dfrac{n}{n-1}$ for $n$.

Multiply both sides by $(n-1)$:

$$x(n-1) = n$$

Remove the brackets:

$$xn - x = n$$

Group the terms containing $n$ on the LHS:

$$xn - n = x$$

$$\therefore \qquad n(x-1) = x$$

Divide both sides by $(x-1)$:

$$n = \frac{x}{x-1}$$

**EXAMPLE 3.10**

Transpose the formula $I_1 = \dfrac{IR_2}{R_1 + R_2}$ to make $R_2$ the subject.

Multiply both sides by $(R_1 + R_2)$:

$$I_1(R_1 + R_2) = IR_2$$

Remove the bracket on the LHS:

$$I_1 R_1 + I_1 R_2 = IR_2$$

Group the terms containing $R_2$ on the RHS:

$$I_1R_1 = IR_2 - I_1R_2$$

Factorise the RHS:

$$I_1R_1 = R_2(I - I_1)$$

Divide both sides by $(I - I_1)$:

$$\frac{I_1R_1}{I - I_1} = R_2$$

$\therefore$ 
$$R_2 = \frac{I_1R_1}{I - I_1}.$$

**EXAMPLE 3.11**

Make $R$ the subject of the formula $\dfrac{1}{R} = \dfrac{1}{R_1} + \dfrac{1}{R_2}$.

The LCM of the RHS is $R_1R_2$

$\therefore$ 
$$\frac{1}{R} = \frac{R_2 + R_1}{R_1R_2}$$

Invert both sides:

$$R = \frac{R_1R_2}{R_2 + R_1}$$

# Exercise 3.1

Transpose the following formulae:

1) $P = \dfrac{RT}{V}$    for $T$

2) $\dfrac{ts}{T} = S$    for $T$

3) $v = u + at$    for $u$

4) $n = p + cr$    for $r$

5) $F = \dfrac{9}{5}C + 32$    for $C$

6) $H = ql + s$    for $l$

7) $S = 115 - 13a$    for $a$

8) $V = \dfrac{2R}{R - r}$    for $r$

9) $C = \dfrac{E}{R + r}$    for $R$

10) $S = \pi r(r + h)$    for $h$

11) $H = ws(T - t)$    for $T$

12) $C = \dfrac{N - n}{2p}$    for $N$

13) $d = \dfrac{2(S-an)}{n(n-l)}$  for $a$

14) $R = \dfrac{1}{1/R_1 + 1/R_2}$  for $R_1$

15) $C = \pi h^2 \left( r - \dfrac{h}{3} \right)$  for $r$

16) $V = \dfrac{2R}{R-r}$  for $R$

17) $P = \dfrac{S(C-F)}{C}$  for $c$

18) $y = \dfrac{2-x}{1+x}$  for $x$

19) $\dfrac{x-p}{q-1} = x$  for $x$

20) $5T = \dfrac{2u-T}{T+u}$  for $u$

# Formulae Containing Roots and Powers

### EXAMPLE 3.12

Transpose $J = I^2 Rt$  for $I$.

Divide both sides by $Rt$:

$$\frac{J}{Rt} = I^2$$

Take the square root of both sides:

$$I = \sqrt{\frac{J}{Rt}}$$

### EXAMPLE 3.13

Transpose $t = 2\pi \sqrt{\dfrac{l}{g}}$  for $l$.

Divide both sides by $2\pi$:

$$\frac{t}{2\pi} = \sqrt{\frac{l}{g}}$$

Square both sides:

$$\left( \frac{t}{2\pi} \right)^2 = \left( \sqrt{\frac{l}{g}} \right)^2$$

or

$$\frac{t^2}{4\pi^2} = \frac{l}{g}$$

Multiply both sides by $g$:

$$l = \frac{t^2 g}{4\pi^2}$$

# Exercise 3.2

Transpose the following formulae:

1) $v = \sqrt{2gh}$   for $h$

2) $A = \pi r^2$   for $r$

3) $E = \frac{1}{2}mv^2$   for $v$

4) $w = 10\sqrt{d}$   for $d$

5) $f = \frac{1}{2\pi}\sqrt{\frac{g}{l}}$   for $l$

6) $a^2 = b^2 + c^2$   for $c$

7) $D = 1.2\sqrt{dL}$   for $L$

8) $x = \sqrt{z^2 - y^2}$   for $y$

9) $P = \frac{f^2 V}{2E}$   for $f$

10) $n = \frac{\pi p r^2}{8vl}$   for $r$

11) $C = 2\sqrt{2hr - h^2}$   for $r$

12) $S = 4\pi r \sqrt{\frac{R^2 + r^2}{2}}$   for $R$

13) $f = \frac{1}{2\pi\sqrt{LC}}$   for $C$

14) $E = mgh + \frac{1}{2}mv^2$   for $v$

15) $V = \pi rl + \pi r^2$   for $l$

16) $\frac{D}{d} = \sqrt{\frac{f+p}{f-p}}$   for $p$

# 4. GRAPHS OF LINEAR EQUATIONS

On reaching the end of this chapter you should be able to:

1. Obtain the standard equation of a straight line $y = mx + c$.
2. Understand the meaning of gradient, m, and intercept, c.
3. Relate linear equations to their graphs.
4. Determine the law of a straight line graph from the gradient and intercept.
5. Determine the law of a straight line graph using the co-ordinates of two points on the line.
6. Plot co-ordinates from a set of experimental data which obey a linear law.
7. Draw the best straight line, by eye, to fit the points.
8. Determine the law from 6 and 7 stating the physical meaning of the gradient where appropriate.
9. Use the equation to calculate other values.

## THE LAW OF A STRAIGHT LINE

In Fig. 4.1, the point B is any point on the line shown and has co-ordinates $x$ and $y$. Point A is where the line cuts the $y$-axis and has co-ordinates $x = 0$ and $y = c$.

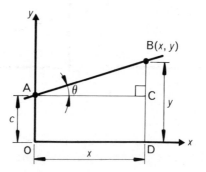

Fig. 4.1

In $\triangle ABC$ $\qquad \dfrac{BC}{AC} = \tan \theta$

∴ $\qquad BC = (\tan \theta) . AC$

32

but also
$$y = BC + CD$$
$$= (\tan\theta).AC + CD$$

$\therefore$
$$\boxed{y = mx + c}$$

This is called the *standard equation*, or *law*, of a straight line

where
$$m = \tan\theta$$

and
$$c = \text{Distance CD} = \text{Distance OA}$$

*m* is called the *gradient of the line*.

*c* is called the *intercept on the y-axis*. Care must be taken as this only applies if the origin (i.e. the point $(0,0)$) is at the intersection of the axes.

In mathematics the gradient of a line is defined as the tangent of the angle that the line makes with the horizontal, and is denoted by the letter *m*.

Hence in Fig. 4.1 the gradient $= m = \tan\theta = \dfrac{BC}{AC}$ *

Fig. 4.2 shows the difference between positive and negative gradients.

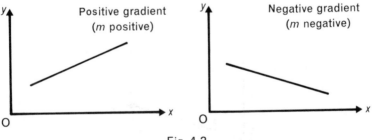

Fig. 4.2

Summarising:

The standard equation, or law, of a straight line is $y = mx + c$

where *m* is the gradient

and *c* is the intercept on the y-axis

*Care should be taken not to confuse this with the gradient given on maps, railways, etc. which is the sine of the angle (not the tangent) — e.g. a railway slope of 1 in 100 is one unit vertically for every 100 units measured along the slope.

# LINEAR EQUATIONS

Equations in which the highest power of the variables is the first, are called equations of the *first degree*. Thus, $y = 3x + 5$, $y = 2 - 7x$ and $y = 0.3x - 0.5$ are all equations of the first degree. All equations of this type give graphs which are straight lines and hence they are often called *linear equations*.

## EXAMPLES OF LINEAR EQUATIONS AND THEIR GRAPHS

You should remember that in the standard equation of a straight line $y = mx + c$ the value of the constant $c$ represents the intercept on the $y$-axis *only* when the origin (i.e. the point $(0, 0)$) is at the intersection of the axes.

Linear equations which are not stated in standard straight line form must be rearranged if they are to be compared with the standard equation $y = mx + c$.

**EXAMPLE 4.1**

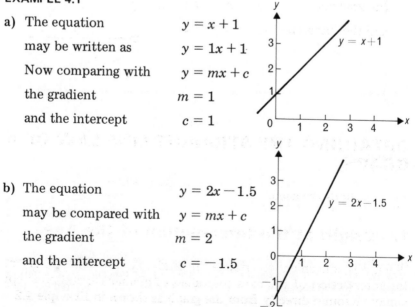

a)  The equation                  $y = x + 1$

    may be written as        $y = 1x + 1$

    Now comparing with      $y = mx + c$

    the gradient                    $m = 1$

    and the intercept            $c = 1$

b)  The equation                  $y = 2x - 1.5$

    may be compared with    $y = mx + c$

    the gradient                    $m = 2$

    and the intercept            $c = -1.5$

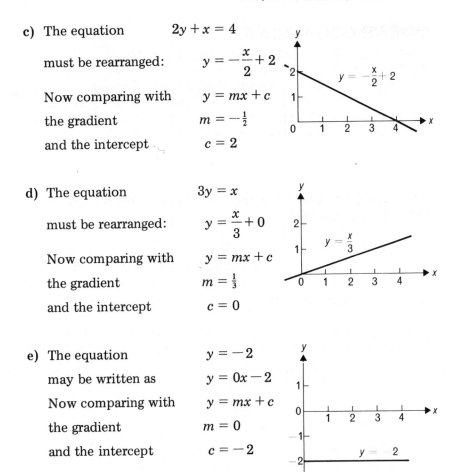

c) The equation $\qquad 2y + x = 4$

    must be rearranged: $\qquad y = -\dfrac{x}{2} + 2$

    Now comparing with $\qquad y = mx + c$

    the gradient $\qquad m = -\frac{1}{2}$

    and the intercept $\qquad c = 2$

d) The equation $\qquad 3y = x$

    must be rearranged: $\qquad y = \dfrac{x}{3} + 0$

    Now comparing with $\qquad y = mx + c$

    the gradient $\qquad m = \frac{1}{3}$

    and the intercept $\qquad c = 0$

e) The equation $\qquad y = -2$

    may be written as $\qquad y = 0x - 2$

    Now comparing with $\qquad y = mx + c$

    the gradient $\qquad m = 0$

    and the intercept $\qquad c = -2$

# OBTAINING THE STRAIGHT LINE LAW OF A GRAPH

Two methods are used:

## 1. Origin at the Intersection of the Axes

When it is convenient to arrange the origin, i.e. the point $(0, 0)$), at the intersection of the axes the values of gradient $m$ and intercept $c$ may be found directly from the graph as shown in Example 4.2.

**EXAMPLE 4.2**

Find the law of the straight line shown in Fig. 4.3.

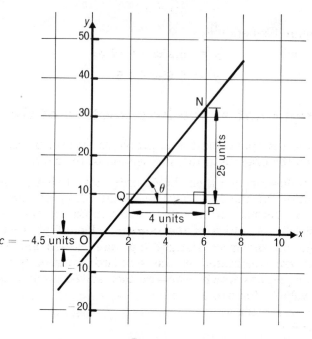

Fig. 4.3

**To find gradient** $m$.    Take any two points Q and N on the line
and construct the right-angled triangle QPN. This triangle should
be of reasonable size, since a small triangle will probably give an
inaccurate result. Note that if we can measure to an accuracy of
1 mm using an ordinary rule, then this error in a length of 20 mm
is much more significant than the same error in a length of 50 mm.

The lengths of NP and QP are then found using the scales of the $x$
and $y$ axes. Direct lengths of these lines as would be obtained
using an ordinary rule, e.g. both in centimetres, must *not* be used
—the scales of the axes must be taken into account.

$$\therefore \qquad \text{Gradient } m = \tan \theta = \frac{NP}{QP} = \frac{25}{4} = 6.25$$

**To find intercept** $c$.    This is measured again using the scale of the
$y$-axis.

$$\therefore \text{ intercept} \qquad\qquad c = -4.5$$

**The law of the straight line**

The standard equation is

$$y = mx + c$$

∴ the required equation is

$$y = 6.25x + (-4.5)$$

i.e.
$$y = 6.25x - 4.5$$

## 2.  Origin not at the Intersection of the Axes

This method is applicable for all problems—it may be used, therefore, when the origin is at the intersection of the axes.

If a point lies on line then the co-ordinates of that point satisfy the equation of the line, e.g. the point $(2, 7)$ lies on the line $y = 2x + 3$ because if $x = 2$ is substituted in the equation, $y = 2 \times 2 + 3 = 7$ which is the correct value of $y$. Two points, which lie on the given straight line, are chosen and their co-ordinates are substituted in the standard equation $y = mx + c$. The two equations which result are then solved simultaneously to find the values of $m$ and $c$.

**EXAMPLE 4.3**

Determine the law of the straight line shown in Fig. 4.4.

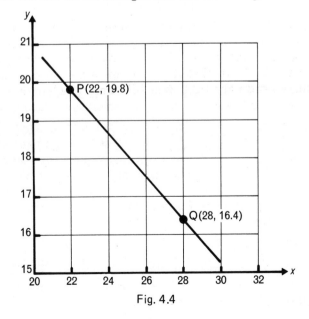

Fig. 4.4

Choose two convenient points P and Q and find their co-ordinates. Again these points should not be close together, but as far apart as conveniently possible. Their co-ordinates are as shown in Fig. 4.4.

Let the equation of the line be $y = mx + c$.

Now P(22, 19.8) lies on the line    $\therefore$  $19.8 = m(22) + c$

and  Q(28, 16.4) lies on the line    $\therefore$  $16.4 = m(28) + c$

To solve these two equations simultaneously we must first eliminate one of the unknowns. In this case $c$ will disappear if the second equation is subtracted from the first, giving

$$19.8 - 16.4 = m(22 - 28)$$

$\therefore$
$$m = \frac{3.4}{-6} = -0.567$$

To find $c$ the value of $m = -0.567$ may be substituted into either of the original equations. Choosing the first equation we get

$$19.8 = -0.567(22) + c$$

$\therefore$
$$c = 19.8 + 0.567(22) = 32.3$$

Hence the required law of the straight line is

$$y = -0.567x + 32.3$$

## GRAPHS OF EXPERIMENTAL DATA

Readings which are obtained as a result of an experiment will usually contain errors owing to inaccurate measurement and other experimental errors. If the points, when plotted, show a trend towards a straight line or a smooth curve, this is usually accepted and the best straight line or curve drawn. In this case the line will not pass through some of the points and an attempt must be made to ensure an even spread of these points above and below the line or the curve.

One of the most important applications of the straight line law is the determination of a law connecting two quantities when values have been obtained from an experiment, as Example 4.4 illustrates.

**EXAMPLE 4.4**

During a test to find how the power of a lathe varied with the depth of cut results were obtained as shown in the table. The speed and feed of the lathe were kept constant during the test.

| Depth of cut, $d$ (mm) | 0.51 | 1.02 | 1.52 | 2.03 | 2.54 | 3.0 |
|---|---|---|---|---|---|---|
| Power, $P$ (W) | 0.89 | 1.04 | 1.14 | 1.32 | 1.43 | 1.55 |

Show that the law connecting $d$ and $P$ is of the form $P = ad + b$ and find the law. Hence find the value of $d$ when $P$ is 1.2 watts.

The standard equation of a straight line is $y = mx + c$. It often happens that the variables are *not* $x$ and $y$. In this example $d$ is used instead of $x$ and is plotted on the horizontal axis, and $P$ is used instead of $y$ and is plotted on the vertical axis.

Similarly the gradient $= a$ instead of $m$, and the intercept on the $y$-axis $= b$ instead of $c$.

On plotting the points (Fig. 4.5) it will be noticed that they deviate slightly from a straight line. Since the data are experimental we must expect errors in observation and measurement and hence a slight deviation from a straight line must be expected.

The points, therefore, approximately follow a straight line and we can say that the equation connecting $P$ and $d$ is of the form $P = ad + b$.

Because the origin is *not* at the intersection of the axes, to find the values of constants $a$ and $b$ we must choose two points *which lie on the line*. These two points must be as far apart as possible in order to obtain maximum accuracy.

In Fig. 4.5 the points P(0.90, 1.00) and Q(2.76, 1.50) have been chosen.

The point P(0.90, 1.00) lies on the line $\therefore$ $1.00 = a(0.90) + b$

and point Q(2.76, 1.50) lies on the line $\therefore$ $1.50 = a(2.76) + b$

Now subtracting the first equation from the second we get

$$1.50 - 1.00 = a(2.76 - 0.90)$$

$$\therefore \qquad a = \frac{0.50}{1.86} = 0.27$$

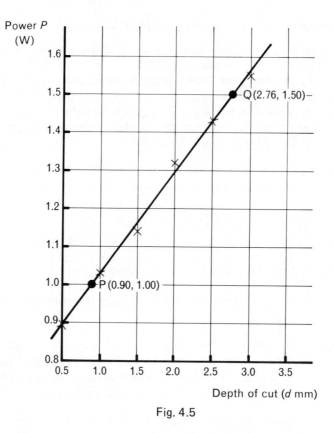

Fig. 4.5

Now substituting the value $a = 0.27$ into the first equation we get

$$1.00 = 0.27(0.90) + b$$

$$\therefore \qquad b = 1.00 - 0.27(0.90) = 0.76$$

Hence the required law of the line is $P = 0.27d + 0.76$

To find $d$ when $P = 1.2$ W. The value $P = 1.2$ is substituted into the equation giving

$$1.2 = 0.27d + 0.76$$

$$d = \frac{1.2 - 0.76}{0.27} = 1.63 \text{ mm}$$

(since all values of $d$ are mm when values of $P$ are watts).

This value of $d$ may be verified by checking the corresponding value of $d$ corresponding to $P = 1.2$ on the straight line in Fig. 4.5.

Any inaccuracies may be due to rounding off calculations to two significant figures, e.g. the value of $m$ is $\dfrac{0.5}{1.86} = 0.269$ if three significant figures are considered. Bearing in mind, however, the experimental errors etc. the rounding off as shown seems reasonable. This question of accuracy is always open to debate, the most dangerous error being to give the calculated results to a far greater accuracy than the original given data.

## EXAMPLE 4.5

Hooke's law states that for an elastic material, up to the limit of proportionality, the stress, $\sigma$, is directly proportional to the strain, $\epsilon$, it produces. In equation form this is $\sigma = E\epsilon$ where the constant $E$ is called the modulus of elasticity of the material.

Find the value of $E$ using the following results obtained in an experiment

| $\sigma$ MN/m$^2$ | 125 | 110 | 95 | 80 |
|---|---|---|---|---|
| $\epsilon$ no units | 0.000 600 | 0.000 522 | 0.000 466 | 0.000 382 |
| $\sigma$ MN/m$^2$ | 63 | 54 | 38 | |
| $\epsilon$ no units | 0.000 367 | 0.000 269 | 0.000 168 | |

In order to compare the equation $\sigma = E\epsilon$ with the standard straight line equation $y = mx + c$ we must plot $\sigma$ on the vertical axis, and $\epsilon$ on the horizontal axis.

Inspection of the values shows that it is convenient to arrange for the origin, i.e. the point $(0,0)$ to be at the intersection of the axes. The graph is shown in Fig. 4.6.

Since the values are the result of an experiment it is unlikely that the points plotted will lie exactly on a straight line. Having arranged for the origin to be at the intersection of the axes we can see from the given equation that the intercept on the $y$-axis is zero. This means that the graph passes through the origin and this helps us in drawing the 'best' straight line to 'fit' the points. Judgement is needed here—for instance the point $(0.000\,367, 63)$ is obviously an incorrect result as it lies well away from the line of the other points, and so we should ignore this point.

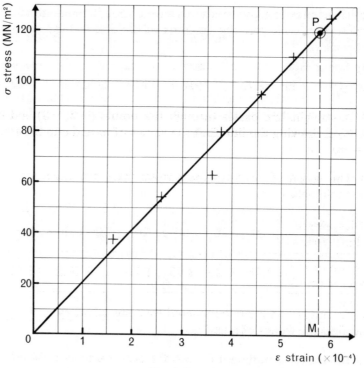

Fig. 4.6

We see also that the gradient $m = E$ on our graph. Thus we may find the value of $E$ by calculating the gradient of the straight line. From the suitable right-angled triangle POM

$$\text{the gradient is given by } \frac{\text{PM}}{\text{OM}} = \frac{120}{0.000\,58} = 207\,000$$

The units of the ratio $\dfrac{\text{PM}}{\text{OM}}$ will be those of PM, i.e. MN/m$^2$, since OM represents strain which has no units (this is because strain is the ratio of two lengths).

Hence the value of the modulus of elasticity of the material $E = 207\,000\,\text{MN/m}^2 = 207\,\text{GN/m}^2$.

## Exercise 4.1

1) Draw the straight line which passes through the points $(4, 7)$ and $(-2, 1)$. Hence find the gradient of the line and its intercept on the y-axis.

**2)** The following equations represent straight lines. Sketch them and find in each case the gradient of the line and the intercept on the $y$-axis.

(a) $y = x + 3$ (b) $y = -3x + 4$

(c) $y = -3.1x - 1.7$ (d) $y = 4.3x - 2.5$.

**3)** A straight line passes through the points $(-2, -3)$ and $(3, 7)$. *Without* drawing the line find the values of $m$ and $c$ in the equation $y = mx + c$.

**4)** The width of keyways for various shaft diameters are given in the table below.

| Diameter of shaft $D$ (mm) | 10 | 20 | 30 | 40 | 50 | 60 | 70 | 80 |
|---|---|---|---|---|---|---|---|---|
| Width of keyway $W$ (mm) | 3.75 | 6.25 | 8.75 | 11.25 | 13.75 | 16.25 | 18.75 | 21.25 |

Show that $D$ and $W$ are connected by a law of the type $W = aD + b$ and find the values of $a$ and $b$.

**5)** During an experiment to find the coefficient of friction between two metallic surfaces the following results were obtained.

| Load $W$ (N) | 10 | 20 | 30 | 40 | 50 | 60 | 70 |
|---|---|---|---|---|---|---|---|
| Friction force $F$ (N) | 1.5 | 4.3 | 7.6 | 10.4 | 13.5 | 15.6 | 18.8 |

Show that $F$ and $W$ are connected by a law of the type $F = aW + b$ and find the values of $a$ and $b$.

**6)** In a test on a certain lifting machine it is found that an effort of 50 N will lift a load of 324 N and that an effort of 70 N will lift a load of 415 N. Assuming that the graph of effort plotted against load is a straight line find the probable load that will be lifted by an effort of 95 N.

**7)** The following results were obtained from an experiment on a set of pulleys. $W$ is the load raised and $E$ is the effort applied. Plot these results and obtain the law connecting $E$ and $W$.

| $W$ (N) | 15 | 20 | 25 | 30 | 35 | 40 | 45 |
|---|---|---|---|---|---|---|---|
| $E$ (N) | 2.3 | 2.7 | 3.2 | 3.8 | 4.3 | 4.7 | 5.3 |

8) During a test with a thermocouple pyrometer the e.m.f. ($E$ millivolts) was measured against the temperature at the hot junction ($t°C$) and the following results were obtained:

| $t$ | 200 | 300 | 400 | 500 | 600 | 700 | 800 | 900 | 1000 |
|---|---|---|---|---|---|---|---|---|---|
| $E$ | 6 | 9.1 | 12.0 | 14.8 | 18.2 | 21.0 | 24.1 | 26.8 | 30.2 |

The law connecting $t$ and $E$ is supposed to be $E = at + b$. Test if this is so and find suitable values for $a$ and $b$.

9) The resistance ($R$ ohms) of a field winding is measured at various temperatures ($t°C$) and the results are recorded in the table below:

| $t (°C)$ | 21 | 26 | 33 | 38 | 47 | 54 | 59 | 66 | 75 |
|---|---|---|---|---|---|---|---|---|---|
| $R$ (ohms) | 109 | 111 | 114 | 116 | 120 | 123 | 125 | 128 | 132 |

If the law connecting $R$ and $t$ is of the form $R = a + bt$ find suitable values of $a$ and $b$.

10) The rate of a spring, $\lambda$, is defined as force per unit extension. Hence for a load, $F$ N producing an extension $x$ mm the law is $F = \lambda x$. Find the value of $\lambda$ in units of N/m using the following values obtained from an experiment:

| $F$ (N) | 20 | 40 | 60 | 80 | 100 | 120 | 140 |
|---|---|---|---|---|---|---|---|
| $x$ (mm) | 37 | 79 | 111 | 156 | 197 | 229 | 270 |

11) A circuit contains a resistor having a fixed resistance of $R$ ohms. The current, $I$ amperes and the potential difference, $V$ volts are related by the expression $V = IR$. Find the value of $R$ given the following results obtained from an experiment:

| $V$ (volts) | 3 | 7 | 11 | 13 | 17 | 20 | 24 | 29 |
|---|---|---|---|---|---|---|---|---|
| $I$ (amperes) | 0.066 | 0.125 | 0.209 | 0.270 | 0.324 | 0.418 | 0.495 | 0.571 |

# GRAPHICAL SOLUTION OF EQUATIONS

On reaching the end of this chapter you should be able to:

1. Solve a pair of two simultaneous linear equations graphically.
2. Determine the roots of a quadratic equation by the intersection of the graph with the x-axis.
3. Solve simultaneously linear and quadratic equations by the intersection of their graphs.
4. Plot the graph of a cubic equation with specified interval and range.
5. Solve a cubic equation using 4.

## GRAPHICAL SOLUTION OF TWO SIMULTANEOUS LINEAR EQUATIONS

Since the solutions we require have to satisfy both the given equations, they will be given by the values of $x$ and $y$ where the graphs of the equations intersect.

The two examples which follow are first solved graphically and then by an alternative theoretical method.

### EXAMPLE 5.1

Solve graphically:

$$y - 2x = 2 \qquad [1]$$
$$3y + x = 20 \qquad [2]$$

Equation [1] may be written as

$$y = 2x + 2$$

and equation [2] may be written as

$$y = -\frac{x}{3} + \frac{20}{3}$$

We will first draw up a table of values. Only three values are necessary in each case since both equations are linear and will have straight line graphs.

45

| $x$ | $-3$ | $0$ | $3$ |
|---|---|---|---|
| $y = 2x + 2$ | $-4$ | $2$ | $8$ |
| $y = -\dfrac{x}{3} + \dfrac{20}{3}$ | $7.7$ | $6.7$ | $5.7$ |

The two graphs are now plotted on the same axes, as shown in Fig. 5.1.

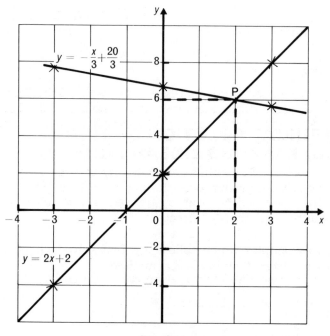

Fig. 5.1

The solutions of the equations will be given by the co-ordinates of the point where the two lines cross (that is, point P in Fig. 4.1). The co-ordinates of P are $x = 2$ and $y = 6$. Hence the solutions are:

$$x = 2 \quad \text{and} \quad y = 6$$

## An Alternative Solution by the Elimination of One Unknown

Multiplying equation [2] by 2 we have

$$6y + 2x = 40 \qquad \qquad [3]$$

Adding equations [1] and [3] we get

$$y - 2x + 6y + 2x = 2 + 40$$

$$\therefore \qquad y = 6$$

Substituting $y = 6$ into equation [1] we have

$$6 - 2x = 2$$

from which $\qquad x = 2$

Thus the solutions are $x = 2$ and $y = 6$ which confirms the graphical results.

**EXAMPLE 5.2**

Two operators are producing the same assemblies. Their total output per week is 220 assemblies. If the ratio of their individual outputs is $6:5$, find the number of assemblies per week that each operator produces.

Let the number of assemblies produced by the faster operator be $x$, and the number produced by the slower operator be $y$.

$$x + y = 220 \qquad [1]$$

and $\qquad \dfrac{x}{y} = \dfrac{6}{5} \qquad [2]$

We now have two simultaneous equations for $x$ and $y$.

Rearranging equation [1] gives

$$y = -x + 220$$

and rearranging equation [2] gives

$$y = \dfrac{5}{6}x$$

We will first draw up a table of values. Only three values are necessary in each case since both equations are linear and will have straight line graphs.

| $x$ | 0 | 100 | 200 |
|---|---|---|---|
| $y = -x + 220$ | 220 | 120 | 20 |
| $y = \dfrac{5}{6}x$ | 0 | 83 | 167 |

The two graphs are now plotted on the same axes, as shown in Fig. 5.2.

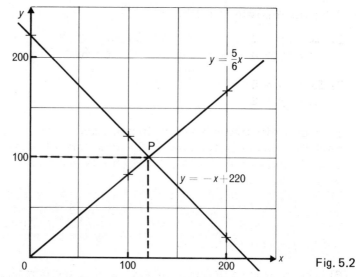

Fig. 5.2

The solutions of the equations will be given by the co-ordinates of the point where the two lines cross (that is, point P in Fig. 5.2). The co-ordinates of P are $x = 120$ and $y = 100$.

Hence the operators produce 120 and 100 assemblies each respectively per week.

## An Alternative Solution Using Substitution

Rearranging equation [2] gives $y = \dfrac{5}{6}x$

We will now substitute this expression for $y$ into equation [1] and obtain

$$x + \frac{5}{6}x = 220$$

from which $\qquad\qquad x = 120$

Now substituting $x = 120$ into equation [1] we have

$$120 + y = 220$$

from which $\qquad\qquad y = 100$

Thus the individual outputs per week of each operator are 120 and 100 assemblies per week. This confirms the result obtained by the graphical method.

# Exercise 5.1

Solve the following equations:

1) $3x + 2y = 7$
   $x + y = 3$

2) $4x - 3y = 1$
   $x + 3y = 19$

3) $2x - 3y = 5$
   $x - 2y = 2$

4) A motorist travels $x$ km at 40 km/h and $y$ km at 50 km/h. The total time taken is $2\frac{1}{2}$ hours. If the time taken to travel $6x$ km at 30 km/h and $4y$ km at 50 km/h is 14 hours find $x$ and $y$.

5) An alloy containing $8 \text{ cm}^3$ of copper and $7 \text{ cm}^3$ of tin has a mass of 122.3 g. A second alloy containing $9 \text{ cm}^3$ of copper and $7 \text{ cm}^3$ of tin has a mass of 131.2 g. Find the densities of copper and tin respectively in $\text{g/cm}^3$.

6) The resistance $R$ ohms of a wire at a temperature of $t\,^\circ\text{C}$ is given by the formula $R = R_0(1 + \alpha t)$ where $R_0$ is the resistance at $0\,^\circ\text{C}$ and $\alpha$ is a constant. The resistance is 35 ohms at a temperature of $80\,^\circ\text{C}$ and 42.5 ohms at a temperature of $140\,^\circ\text{C}$. Find $R_0$ and $\alpha$. Hence find the resistance when the temperature is $50\,^\circ\text{C}$.

7) If 100 m of wire and 8 plugs cost £12.40 and 150 m of wire and 10 plugs cost £18. Find the cost of 1 m of wire and the cost of a plug.

8) A heating installation for one house consists of 5 radiators and 4 convector heaters and the cost, including labour, is £1040. In a second house 6 radiators and 7 convector heaters are used and the cost, including labour, is £1538. In each house the installation costs are £200. Find the cost of a radiator and the cost of a convector heater.

9) If 100 m of tubing and 8 elbow fittings cost £100.00, and 150 m of tubing and 10 elbow fittings cost £147.50, find the cost of 1 m of tubing and the cost of an elbow fitting.

10) For a builder's winch it is found that the effort $E$ newtons and the load $W$ newtons are connected by the equation $E = aW + b$. An effort of 90 N lifts a load of 100 N whilst an effort of 130 N lifts a load of 200 N. Find the values of $a$ and $b$ and hence determine the effort required to lift a load of 300 N.

**11)** A penalty clause states that a contractor will forfeit a certain sum of money for each day that he is late in completing a contract (i.e. the contractor gets paid the value of the original contract less any sum forfeit). If he is 6 days late he receives £5000 and if he is 14 days late he receives £3000. Find the amount of the daily forfeit and determine the value of the original contract.

**12)** The total cost of equipping two laboratories, A and B, is £30 000. If laboratory B costs £2000 more than laboratory A find the cost of the equipment for each of them.

**13)** $x$ kg of a chemical and $2y$ kg of a second chemical cost £90. If the mass ratio of the chemicals is inverted the cost is only £60. Assuming that each chemical costs £5 per kg find the masses $x$ and $y$.

**14)** The sum of the ages of two installations is 46 months. The modern version is 10 months younger than the original. Calculate their present ages.

**15)** A company's annual net profit of £8800 is divided amongst the two partners in the ratio $x : 2y$. If the first shareholder receives £2000 more than the other, find the values of $x$ and $y$, and hence the respective shares of the profit.

**16)** The cost of a bacteria culture $b$ is £0.60 and of a bacteria culture $B$ is £0.90. A combination of several samples of each culture costs £9. Inflation raises the cost of each sample by 15 p and the cost of the combination to £10.65. Find the number of samples of each bacteria cultured.

# GRAPHICAL SOLUTION OF QUADRATIC EQUATIONS

An equation in which the highest power of the unknown is two, and containing no higher powers of the unknown, is called a quadratic equation. It is also known as an equation of the *second degree*. Thus

$$x^2 - 9 = 0 \qquad 2.5x^2 - 3.1x - 2 = 0$$
$$x^2 + 2x - 8 = 0 \qquad 2x^2 - 5x = 0$$

are all examples of quadratic equations.

Quadratic equations may be solved by plotting graphs. This method is explained fully by the examples which follow.

## EXAMPLE 5.3

Plot the graph of $y = 3x^2 + 10x - 8$ between $x = -6$ and $x = +4$. Hence solve the equation $3x^2 + 10x - 8 = 0$.

A table can be drawn up as follows giving values of $y$ for the chosen values of $x$.

| $x$ | $-6$ | $-5$ | $-4$ | $-3$ | $-2$ | $-1$ |
|---|---|---|---|---|---|---|
| $3x^2$ | 108 | 75 | 48 | 27 | 12 | 3 |
| $10x$ | $-60$ | $-50$ | $-40$ | $-30$ | $-20$ | $-10$ |
| $-8$ | $-8$ | $-8$ | $-8$ | $-8$ | $-8$ | $-8$ |
| $y$ | 40 | 17 | 0 | $-11$ | $-16$ | $-15$ |

| $x$ | 0 | 1 | 2 | 3 | 4 |
|---|---|---|---|---|---|
| $3x^2$ | 0 | 3 | 12 | 27 | 48 |
| $10x$ | 0 | 10 | 20 | 30 | 40 |
| $-8$ | $-8$ | $-8$ | $-8$ | $-8$ | $-8$ |
| $y$ | $-8$ | 5 | 24 | 49 | 80 |

The graph of $y = 3x^2 + 10x - 8$ is shown plotted in Fig. 5.3.

To solve the equation $3x^2 + 10x - 8 = 0$ we have to find the value of $x$ when $y = 0$, that is, the value of $x$ where the graph cuts the $x$-axis.

These are the points A and B in Fig. 5.3.

Hence the solutions of $3x^2 + 10x - 8 = 0$ are

$$x = -4 \quad \text{and} \quad x = 0.7$$

The accuracy of the results obtained by this method will depend on the scales chosen. The value $x = -4$ is exact as this value was taken when drawing up the table of values, and gave $y = 0$. The value $x = 0.7$ is as accurate as may be read from the scale chosen for the $x$-axis.

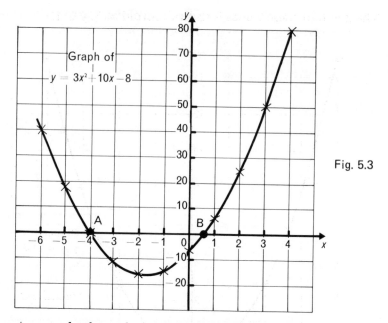

Fig. 5.3

The next example shows how to obtain a more accurate result by plotting a portion of the graph with larger scales.

### EXAMPLE 5.4

Find the roots of the equation $x^2 - 1.3 = 0$ by drawing a suitable graph.

Using the same method as in Example 5.3 we need to plot a graph of $y = x^2 - 1.3$ and find where it cuts the $x$-axis.

We have not been given a range of values of $x$ between which the curve should be plotted and so we must make our own choice.

A good method is to try first a range from $x = -4$ to $x = +4$. If only five values of $y$ are calculated for values of $x$ of $-4, -2,$ $0, +1$ and $+2$, we shall not have wasted much time if these values are not required—in any case we shall learn from this trial and be able to make a better choice at the next attempt.

The first table of values is as follows:

| $x$ | $-4$ | $-2$ | $0$ | $2$ | $4$ |
|---|---|---|---|---|---|
| $x^2$ | $16$ | $4$ | $0$ | $4$ | $16$ |
| $-1.3$ | $-1.3$ | $-1.3$ | $-1.3$ | $-1.3$ | $-1.3$ |
| $y$ | $14.7$ | $2.7$ | $-1.3$ | $2.7$ | $14.7$ |

The graph of these values is shown plotted in Fig. 5.4.

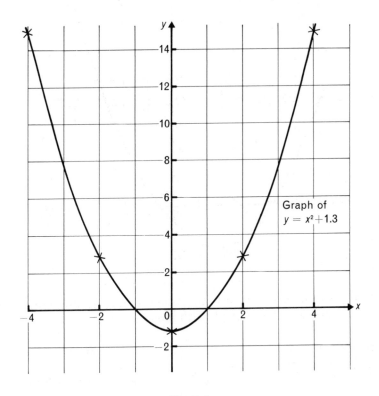

Graph of
$y = x^2 + 1.3$

Fig. 5.4

The approximate values of $x$ where the curve cuts the $x$-axis are
$-1$ and $+1$ (Fig. 5.4). For more accurate results we must plot the
portion of the curve where it cuts the $x$-axis to a larger scale. We
can see, however, both from the table of values and the graph, that
the graph is symmetrical about the $y$-axis—so we need only plot
one half. We will choose the portion to the right of the $y$-axis and
draw up a table of values from $x = 0.7$ to $x = 1.3$

| $x$ | 0.7 | 0.8 | 0.9 | 1.0 | 1.1 | 1.2 | 1.3 |
|---|---|---|---|---|---|---|---|
| $x^2$ | 0.49 | 0.64 | 0.81 | 1.00 | 1.21 | 1.44 | 1.69 |
| $-1.3$ | $-1.3$ | $-1.3$ | $-1.3$ | $-1.3$ | $-1.3$ | $-1.3$ | $-1.3$ |
| $y$ | $-0.81$ | $-0.66$ | $-0.49$ | $-0.30$ | $-0.09$ | 0.14 | 0.39 |

These values are shown plotted in Fig. 5.5

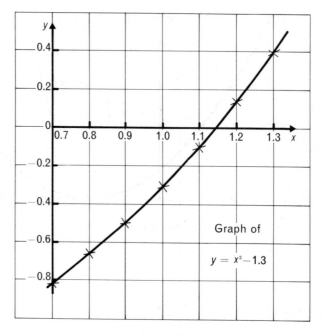

Fig. 5.5

The graph cuts the $x$-axis where $x = 1.14$ (Fig. 5.5) and we must not forget the other value of $x$ where the curve cuts the $x$-axis to the left of the $y$-axis. This will be where $x = -1.14$ since the curve is symmetrical about the $y$-axis.

Hence the solutions of $x^2 - 1.3 = 0$ are

$$x = 1.14 \quad \text{and} \quad x = -1.14$$

**EXAMPLE 5.5**

Solve the equation $x^2 - 4x + 4 = 0$

We shall plot the graph of $y = x^2 - 4x + 4$ and find where it cuts the $x$-axis.

The graph is shown plotted in Fig. 5.6. In this case the curve does not actually cut the $x$-axis but touches it at the point where $x = 2$. Another way of looking at it is to say that the curve 'cuts' the $x$-axis at two points which lie on top of each other. The two points coincide and they are said to be coincident points. The roots are called repeated roots.

The only solution to $x^2 - 4x + 4 = 0$ is, therefore, $x = 2$

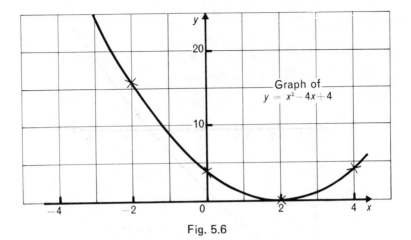

Fig. 5.6

## EXAMPLE 5.6

Solve the equation $x^2 + x + 3 = 0$

We shall plot the graph of $y = x^2 + x + 3$ and find where it cuts the $x$-axis.

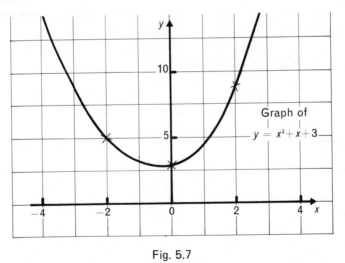

Fig. 5.7

The graph is shown plotted in Fig. 5.7. We can see that the curve does not cut the $x$-axis at all. This means there are no roots—in theory there are roots but they are complex or imaginary and have no arithmetical value.

## Exercise 5.2

By plotting suitable graphs solve the following equations:

1) $x^2 - 7x + 12 = 0$    (plot between $x = 0$ and $x = 6$)

2) $x^2 + 16 = 8x$    (plot between $x = 1$ and $x = 7$)

3) $x^2 - 9 = 0$    (plot between $x = -4$ and $x = 4$)

4) $x^2 + 2x - 15 = 0$    5) $3x^2 - 23x + 14 = 0$

6) $2x^2 + 13x + 15 = 0$    7) $x^2 - 2x - 1 = 0$

8) $3x^2 - 7x + 1 = 0$    9) $9x^2 - 5 = 0$

# GRAPHICAL SOLUTION OF SIMULTANEOUS LINEAR AND QUADRATIC EQUATIONS

Since the solutions we require have to satisfy both the given equations they will be given by the values of $x$ and $y$ where the graphs of the equations intersect.

### EXAMPLE 5.7

Solve simultaneously the equations:

$$y = x^2 + 3x - 4$$

and    $$y = 2x + 4$$

We must first draw up tables of values, and will use the range $x = -4$ to $x = +4$:

| $x$ | $-4$ | $-2$ | $0$ | $2$ | $4$ |
|---|---|---|---|---|---|
| $x^2$<br>$+ 3x$<br>$- 4$ | $16$<br>$-12$<br>$-4$ | $4$<br>$-6$<br>$-4$ | $0$<br>$0$<br>$-4$ | $4$<br>$6$<br>$-4$ | $16$<br>$12$<br>$-4$ |
| $y = x^2 + 3x - 4$ | $0$ | $-6$ | $-4$ | $6$ | $24$ |
| $2x$<br>$+ 4$ | $-8$<br>$4$ | $-4$<br>$4$ | $0$<br>$4$ | $4$<br>$4$ | $8$<br>$4$ |
| $y = 2x + 4$ | $-4$ | $0$ | $4$ | $8$ | $12$ |

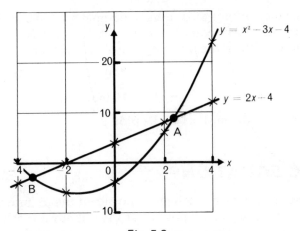

Fig. 5.8

The two graphs are shown plotted on the same axes in Fig. 5.8 and they intersect at the points A and B. Values of the $x$ and $y$ co-ordinates at these points will give the solutions of the given equations. We must be careful not to try to read the values too accurately, as the graphs have been plotted using only five values of $x$. In this case, even values to the first place of decimals cannot be guaranteed.

If more accurate answers are required then we must plot the portions of the graph containing points A and B using more values of $x$ and also much larger scales.

Hence the required solutions are

| $x$ | 2.4 | $-3.4$ |
|---|---|---|
| $y$ | 8.7 | $-2.7$ |

# Exercise 5.3

Solve simultaneously:

1) $y = x^2 - 2x - 2$
   $x - y + 2 = 0$

2) $y = x^2 - x + 5$
   $y = 2x + 5$

3) $y = 5x^2 + x - 3$
   $y = 5x - 2$

4) $y = 2x^2 - 2.3x + 1$
   $y = 3x - 0.25$

5) A rectangular plot of land has a perimeter of 280 m. The length of a diagonal drawn corner to corner is 100 m. If $x$ and $y$ are the length and width of the plot respectively show that:

$$x + y = 140 \qquad [1]$$

and $\qquad\qquad x^2 + y^2 = 10\,000 \qquad [2]$

Hence find the dimensions of the plot.

# CUBIC EQUATIONS

An equation in which the highest power of the unknown is three, and containing no higher powers of the unknown is called a cubic equation. It is also known as an equation of the *third degree*. Thus

$$x^3 - 37 = 0$$
$$x^3 + 2x^2 + 1 = 0$$
$$3x^3 - 4x - 13 = 0$$
$$x^3 + x^2 + 7x - 10 = 0$$

are all examples of cubic equations.

The algebraic method of solving cubic equations is difficult and we shall solve cubic equations by a graphical method similar to that used for solving quadratic equations. This, in fact, is the usual way technicians solve cubic equations when they arise from practical problems.

### EXAMPLE 5.8

Plot the graph of $y = x^3 - 1.5x^2 - 8.5x + 4.5$ from $x = -4$ to $x = +4$ at 1 unit intervals, and use the graph to solve the cubic equation $x^3 - 1.5x^2 - 8.5x + 4.5 = 0$

A table can be drawn up as follows giving values of $y$ for the chosen values of $x$:

| $x$ | $-4$ | $-3$ | $-2$ | $-1$ | $0$ |
|---|---|---|---|---|---|
| $x^3$ | $-64$ | $-27$ | $-8$ | $-1$ | $0$ |
| $-1.5x^2$ | $-24$ | $-13.5$ | $-6$ | $-1.5$ | $0$ |
| $-8.5x$ | $34$ | $25.5$ | $17$ | $8.5$ | $0$ |
| $+4.5$ | $4.5$ | $4.5$ | $4.5$ | $4.5$ | $4.5$ |
| $y$ | $-49.5$ | $-10.5$ | $7.5$ | $10.5$ | $4.5$ |

| $x$ | 1 | 2 | 3 | 4 |
|---|---|---|---|---|
| $x^3$ $-1.5x^2$ $-8.5x$ $+4.5$ | 1 $-1.5$ $-8.5$ 4.5 | 8 $-6$ $-17$ 4.5 | 27 $-13.5$ $-25.5$ 4.5 | 64 $-24$ $-34$ 4.5 |
| $y$ | $-4.5$ | $-10.5$ | $-7.5$ | 10.5 |

The graph is shown plotted in Fig. 5.9. To solve the cubic equation we have to find the values of $x$ when $y = 0$, that is, the value of $x$ where the graph cuts the $x$-axis.

Hence the required solutions of $x^3 - 1.5x^2 - 8.5x + 4.5 = 0$ are

$$x = -2.5, \quad x = 0.5 \quad \text{and} \quad x = 3.5$$

We must remember that these values are only approximate, since the values of the first decimal place cannot be guaranteed. As in previous examples, if more accurate answers are required then we must plot the portions of the graph containing the points of intersection using more values of $x$ and also much larger scales.

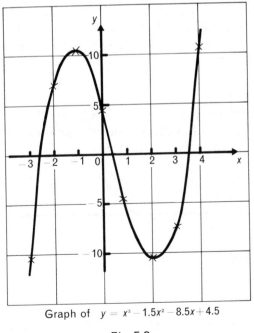

Graph of $y = x^3 - 1.5x^2 - 8.5x + 4.5$

Fig. 5.9

**EXAMPLE 5.9**

A domed roof is in the form of a cap of a sphere. Its base radius is 10 m. If the height of the dome is $h$ metres and the volume of air space under the dome is $1525\,\text{m}^3$. It can be shown that

$$h^3 + 300h - 2912 = 0$$

Plot the graph of $y = h^3 + 300h - 2912$ for values of $h$ between 4 and 12 and hence find the value of $h$.

A table is drawn up giving values of $y$ corresponding to the chosen values of $h$:

| $h$ | 4 | 5 | 6 | 7 | 8 |
|---|---|---|---|---|---|
| $h^3$ 300$h$ $-2912$ | 64 1200 $-2912$ | 125 1500 $-2912$ | 216 1800 $-2912$ | 343 2100 $-2912$ | 512 2400 $-2912$ |
| $y$ | $-1648$ | $-1287$ | $-896$ | $-469$ | 0 |

| $h$ | 9 | 10 | 11 | 12 |
|---|---|---|---|---|
| $h^3$ 300$h$ $-2912$ | 729 2700 $-2912$ | 1000 3000 $-2912$ | 1331 3300 $-2912$ | 1728 3600 $-2912$ |
| $y$ | 517 | 1088 | 1719 | 2416 |

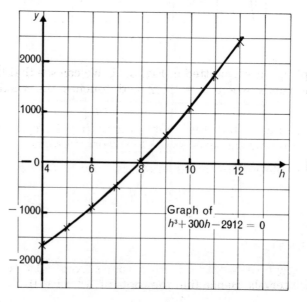

Fig. 5.10

Graph of $h^3 + 300h - 2912 = 0$

The graph is drawn in Fig. 5.10 and the solution of the equation $h^3 + 300h - 2912 = 0$ is the value of $h$ where the curve cuts the horizontal axis. Hence $h = 8$ is the solution. Note that this solution is exact since the table shows that when $h = 8$, $y = 0$

**EXAMPLE 5.10**

Plot the graph of $y = 5x^3 - 9x^2 + 3x + 1$ from $x = -0.4$ to $x = +1.4$ at intervals of 0.2 units, and hence find the values of the roots of the equation $5x^3 - 9x^2 + 3x + 1 = 0$

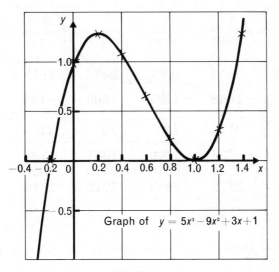

Fig. 5.11

The graph is shown plotted in Fig. 5.11 We can see that the curve cuts the $x$-axis where $x = -0.2$ and touches the $x$-axis where $x = 1$

As in Example 5.5 the point where $x = 1$ represents two coincident points and gives rise to two repeated roots.

Hence the solutions of the equation $5x^3 - 9x^2 + 3x + 1 = 0$ are

$$x = -0.2 \quad \text{and} \quad x = 1$$

**EXAMPLE 5.11**

Plot the graph of $y = x^3 - 1$ for $x$ values from $-1.0$ to $+1.5$ at half unit intervals. Hence find the roots of $x^3 - 1 = 0$

The graph is shown plotted in Fig. 5.12 and it can be seen that the curve only cuts the x-axis at one point which gives the value of the only real root. The other two solutions are complex or imaginary and have no arithmetical meaning.

The only real solution of $x^3 - 1 = 0$ is, therefore, $x = 1$

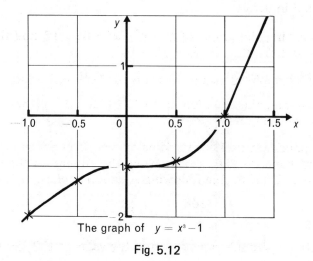

The graph of $y = x^3 - 1$

Fig. 5.12

# Exercise 5.4

1) Plot the graph of $y = x^3 - 4x^2 - x + 4$ from $x = -2$ to $x = 5$ at one unit intervals. Hence solve the equation $x^3 - 4x^2 - x + 4 = 0$.

2) Plot the graph of $y = 3x^3 - 3.4x^2 - 6.4x + 2.4$ using values of $x$ from $-2$ to $+3$ at half unit intervals. Hence find the roots of the cubic equation $3x^3 - 3.4x^2 - 6.4x + 2.4 = 0$.

3) Plot the graph of $y = x^3 + x^2 + x - 3$ from $x = -3$ to $x = +3$ at one unit intervals. Hence show that the cubic equation $x^3 + x^2 + x - 3 = 0$ has only one real root, and find its value.

4) Plot the graph of $y = x^3 - x^2 - x + 1$ taking values of $x$ from $-2$ to $+2$ at half unit intervals. Hence find the roots of the equation $x^3 - x^2 - x + 1 = 1$.

5) Plot the graph of $y = 4x^3 - 15x^2 + 7x + 6$ from $x = -1$ to $x = 3.5$ at intervals of one half unit. Use this graph to solve the cubic equation $4x^3 - 15x^2 + 7x + 6 = 0$.

**6)** Plot the graph of $y = x^3 - x^2 - x - 2$ for values of $x$ from $-4$ to $+4$ at one unit intervals, and hence find the roots of the equation $x^3 - x^2 - x - 2 = 0$.

**7)** Find the roots of the cubic equation $3x^3 + 4x^2 - 12x + 5 = 0$ by plotting a suitable graph taking values of $x$ from $-3.5$ to $+1.5$ at half unit intervals.

**8)** By plotting the graph of $y = x^3 - x^2 - 8x + 12$ find the roots of the equation $x^3 - x^2 - 8x + 12 = 0$.

**9)** Find by graphical means the roots of $7x^3 - 6x^2 - 18x + 4 = 0$.

**10)** Show that the equation $2x^3 + 3x^2 + 2x + 1 = 0$ has only one real root, and find its value.

**11)** A spherical vessel has a radius of 8 m. It contains liquid to a height of $h$ metres (Fig. 5.13). When the volume of the liquid in the vessel is $1.056 \text{ m}^3$ the following equation applies:

$$\frac{3168}{\pi} = h^2(24 - h)$$

By plotting a suitable equation find the value of $h$ ($h$ lies between 3 and 8).

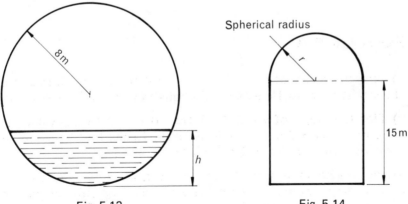

Fig. 5.13                    Fig. 5.14

**12)** A pressure vessel (Fig. 5.14) has a capacity of $594 \text{ m}^3$. The equation from which $r$ may be found is

$$2r^3 + 45r^2 - 567 = 0$$

By drawing a suitable graph find the value of $r$ ($r$ lies between 0 m and 5 m).

**13)** A domed roof is in the form of a segment of a sphere. If the volume of air space under the dome is $497 \text{ m}^3$ and the radius of the sphere is 8 m, then

$$h^3 - 24h^2 + 475 = 0$$

Plot a suitable graph to find $h$, the height of the dome, given that its value is between 4 m and 6 m.

**14)** A cement silo is in the form of a frustum of a cone with a height equal to the larger radius of the frustum. The following equation then applies:

$$R^3 + 2R^2 + 4R - 57 = 0$$

Find the value of $R$, the height of the silo, by plotting a suitable graph.

# RADIAN MEASURE

---

---

## RADIAN MEASURE

We have seen that an angle is usually measured in degrees but there is another way of measuring an angle. In this, the unit is known as the radian (abbreviation rad).

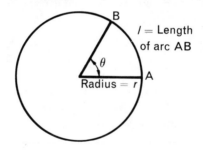

Fig. 6.1

Referring to Fig. 6.1 gives

$$\text{Angle in radians} = \frac{\text{Length of arc}}{\text{Radius of circle}}$$

$$\theta \text{ radians} = \frac{l}{r}$$

or
$$l = r\theta$$

# RELATION BETWEEN RADIANS AND DEGREES

If we make the arc AB equal to a semi-circle then

$$\text{Length of arc} = \pi r$$

and $$\text{Angle in radians} = \frac{\pi r}{r} = \pi$$

Now the angle subtended by a semi-circle $= 180°$

Therefore $$\pi \text{ radians} = 180°$$

or $$1 \text{ radian} = \frac{180°}{\pi} = 57.3°$$

Thus to convert from degrees to radians

$$\boxed{\theta° = \frac{\pi\theta}{180} \text{ radians}}$$

Thus $$30° = \frac{\pi(30)}{180} \text{ rad} = \frac{\pi}{6} \text{ rad}$$

$$90° = \frac{\pi}{2} \text{ rad} \qquad\qquad 180° = \pi \text{ rad}$$

$$45° = \frac{\pi}{4} \text{ rad} \qquad\qquad 270° = \frac{3\pi}{2} \text{ rad}$$

$$60° = \frac{\pi}{3} \text{ rad} \qquad\qquad 360° = 2\pi \text{ rad}$$

To convert from radians to degrees

$$\boxed{\theta \text{ radians} = \left(\frac{180}{\pi} \times \theta\right)°}$$

### EXAMPLE 6.1

Convert $29°37'29''$ to radians stating the answer correct to 4 significant figures.

The first step is to convert the given angle into degrees and decimals of a degree.

$$29°37'29'' = 29 + \frac{37}{60} + \frac{29}{3600} = 29.625°$$

$$= \frac{\pi \times 29.625}{180} = 0.5171 \text{ radians}$$

Many scientific calculators will convert degrees, minutes and seconds into decimal degrees, and vice versa, using special keys — instructions for use of these keys will be given in the accompanying booklet.

However, the sequence given below will perform the calculation given in the solution shown above.

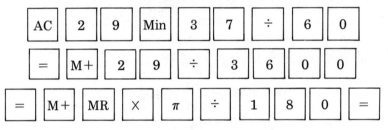

giving an answer of 0.5171 correct to 4 significant figures.

### EXAMPLE 6.2

Convert 0.089 35 radians into degrees, minutes and seconds.

$$0.089\,35 \text{ radians} = \frac{0.089\,35 \times 180}{\pi} = 5.1194° = 5°7'10''$$

For calculators without decimal degree conversion facility the following sequence may be used—this is a reverse of the sequence used in the previous example.

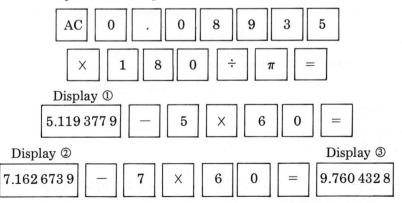

In this sequence it is necessary to record three results as they appear. The whole number, namely 5, in display ① is the number of degrees. The whole number 5 is then subtracted to leave the decimal part which is then multiplied by 60. The whole number, namely 7, in display ② is the number of minutes. The whole number, namely 7 is now subtracted to leave the decimal part which is then multiplied by 60 to give display ③—this figure is the number of seconds.

Thus the result is $5°7'10''$ to the nearest second.

## THE AREA OF A SECTOR

The area of a circle $= \pi r^2$.

So, by proportion, referring to Fig. 6.2 gives

$$\text{Area of sector} = \pi r^2 \times \frac{\theta}{2\pi} = \tfrac{1}{2} r^2 \theta$$

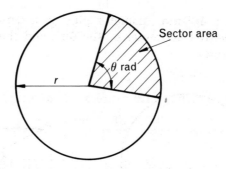

Fig. 6.2

**EXAMPLE 6.3**

Find the angle of a sector of radius 35 mm and area 1020 mm²

Now        Area of sector $= \tfrac{1}{2} r^2 \theta$

and substituting the given values of

$$\text{Area} = 1020 \text{ mm}^2 \quad \text{and} \quad r = 35 \text{ mm}$$

we have        $1020 = \tfrac{1}{2}(35)^2 \theta$

from which        $\theta = \dfrac{1020 \times 2}{35^2} = 1.67 \text{ rad}$

$$= \frac{180 \times 1.67}{\pi} = 95.7°$$

*Summary*

| Length of arc of a sector $= r\theta$ or $2\pi r\left(\dfrac{\theta^{\circ}}{360}\right)$ |
|---|
| Area of a sector $= \frac{1}{2}r^2\theta$ or $\pi r^2\left(\dfrac{\theta^{\circ}}{360}\right)$ |

### EXAMPLE 6.4

Calculate a) the length of arc of a circle whose radius is 8 m and which subtends an angle of 56° at the centre, and b) the area of the sector so formed.

a) Length of arc $= 2\pi r \times \dfrac{\theta^{\circ}}{360} = 2 \times \pi \times 8 \times \dfrac{56}{360} = 7.82\,\text{m}$

b) Area of sector $= \pi r^2 \times \dfrac{\theta^{\circ}}{360} = \pi \times 8^2 \times \dfrac{56}{360} = 31.28\,\text{m}^2$

### EXAMPLE 6.5

Water flows in a 400 mm diameter pipe to a depth of 300 mm. Calculate the wetted perimeter of the pipe and the area of cross-section of the water.

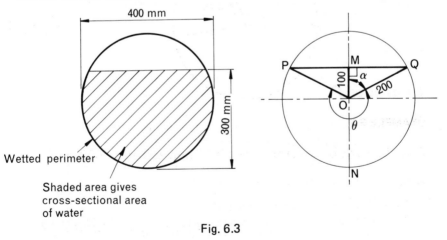

Fig. 6.3

From Fig. 6.3

the right-angled triangle MQO

$$\cos\alpha = \frac{OM}{OQ} = \frac{100}{200} = 0.5$$

$$\therefore \qquad \alpha = 60^{\circ}$$

Also $\qquad \sin \alpha = \dfrac{MQ}{OQ}$

$\therefore \qquad MQ = OQ \sin \alpha = 200 \sin 60° = 173.2 \, mm$

Now $\qquad \theta + 2\alpha = 360°$

$\therefore \qquad \theta = 360° - 2(60°) = 240°$

Thus

Wetted perimeter $= Arc \, PNQ$

$$= 2\pi r \left(\dfrac{\theta}{360}\right) = 2\pi(200)\left(\dfrac{240}{360}\right) = 838 \, mm$$

Also

$$\left(\begin{array}{c} \text{Cross-sectional} \\ \text{area of water} \end{array}\right) = \left(\begin{array}{c} \text{Area of} \\ \text{sector PNQ} \end{array}\right) + \left(\begin{array}{c} \text{Area of} \\ \text{triangle POQ} \end{array}\right)$$

$$= \pi r^2 \left(\dfrac{\theta}{360}\right) + \tfrac{1}{2}(PQ)(MO)$$

$$= \pi(200)^2 \left(\dfrac{240}{360}\right) + \tfrac{1}{2}(2 \times 173.2)(100)$$

$$= 83\,780 + 17\,320$$

$$= 101\,000 \, mm^2$$

# Exercise 6.1

1) Convert the following angles to radians stating the answers correct to 4 significant figures:

(a) 35°  (b) 83°28′  (c) 19°17′32″  (d) 43°39′49″

2) Convert the following angles to degrees, minutes and seconds correct to the nearest second:

(a) 0.1732 radians  (b) 1.5632 radians  (c) 0.0783 radians

3) If $r$ is the radius and $\theta$ is the angle subtended by an arc, find the length of arc when:

(a) $r = 2\,m, \quad \theta = 30°$  (b) $r = 34\,mm, \quad \theta = 38°40′$

4) If $l$ is the length of an arc, $r$ is the radius and $\theta$ the angle subtended by the arc, find $\theta$ when:

(a) $l = 9.4\,m, \quad r = 4.5\,m$  (b) $l = 14\,mm, \quad r = 79\,mm$

5) If an arc 70 mm long subtends an angle of 45° at the centre, what is the radius of the circle?

6) Find the areas of the following sectors of circles:
(a)  radius 3 m, angle of sector 60°
(b)  radius 27 mm, angle of sector 79°45′
(c)  radius 78 mm, angle of sector 143°42′

7) Calculate the area of the part shaded in Fig. 6.4.

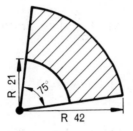

Fig. 6.4

8) A chord 26 mm is drawn in a circle of 35 mm diameter. What are the lengths of arcs into which the circumference is divided?

9) The radius of a circle is 60 mm. A chord is drawn 40 mm from the centre. Find the area of the minor segment.

10) In a circle of radius 30 mm a chord is drawn which subtends an angle of 80° at the centre. What is the area of the minor segment?

11) A flat is machined on a circular bar of 15 mm diameter, the maximum depth of cut being 2 mm. Find the area of the cross-section of the finished bar.

12) Water flows in a 300 mm diameter drain to a depth of 100 mm. Calculate the wetted perimeter of the drain and the area of cross-section of the water.

13) In marking out the plan of part of a building, a line 8 m long is pegged down at one end. Then with the line held horizontal and taut, the free end is swung through an angle of 57°. Calculate the distance moved by the free end of the line and determine the area swept out.

**14)** Find the area of brickwork necessary to fill in the tympanum of the segmental arch shown in Fig. 6.5.

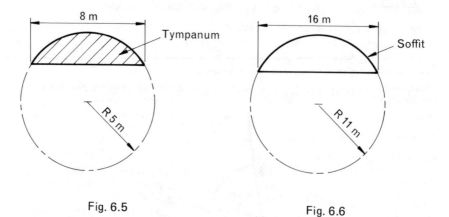

Fig. 6.5                                   Fig. 6.6

**15)** Fig. 6.6 shows a segmental arch for a bridge. Calculate the length of the soffit of the arch.

# 7. TRIGONOMETRY

After reaching the end of this chapter you should be able to:

1. Understand what is meant by a trigonometrical identity.

2. Show that
$$\tan A = \frac{\sin A}{\cos A} \text{ and } \sin^2 A + \cos^2 A = 1$$
using a right-angled triangle.

3. Define the reciprocal ratios: cosecant, secant and cotangent.

4. Find the values of trigonometrical ratios of angles between $0°$ and $360°$.

5. Sketch the sine and cosine waveforms over one complete cycle from the projections of a rotating unit radius.

6. Sketch the tangent curve using the identity
$$\tan \theta = \frac{\sin \theta}{\cos \theta}$$

7. Describe the periodic properties of the sine, cosine and tangent curves.

8. State the sine rule for a suitably labelled triangle in the form
$$\frac{a}{\sin A} = \frac{b}{\sin B} = \frac{c}{\sin C} = D$$

9. Recognise conditions under which the sine rule may be used, and apply it to the solution of practical problems.

10. State the cosine rule for a suitably labelled triangle in the form
$$a^2 = b^2 + c^2 - 2bc \cos A$$

11. Recognise conditions under which the cosine rule may be used, and apply it to the solution of practical problems.

## THE TRIGONOMETRICAL RATIOS OF A RIGHT-ANGLED TRIANGLE

The definitions of the three trigonometrical ratios have been given in *Mathematics for Technicians—New Level I*. However, for reference, they are given again below. Refer to Fig. 7.1.

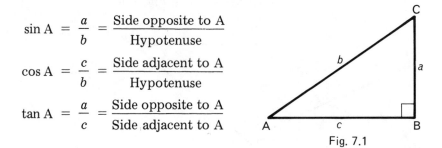

$$\sin A = \frac{a}{b} = \frac{\text{Side opposite to A}}{\text{Hypotenuse}}$$

$$\cos A = \frac{c}{b} = \frac{\text{Side adjacent to A}}{\text{Hypotenuse}}$$

$$\tan A = \frac{a}{c} = \frac{\text{Side opposite to A}}{\text{Side adjacent to A}}$$

Fig. 7.1

**EXAMPLE 7.1**

Find the sides marked $x$ in Figs. 7.2, 7.3 and 7.4.

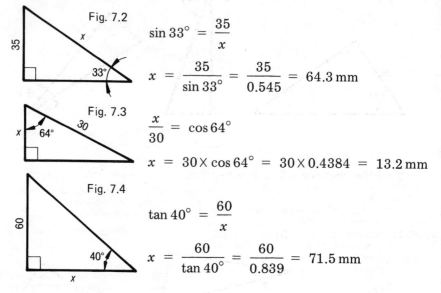

Fig. 7.2

$$\sin 33° = \frac{35}{x}$$

$$x = \frac{35}{\sin 33°} = \frac{35}{0.545} = 64.3\,\text{mm}$$

Fig. 7.3

$$\frac{x}{30} = \cos 64°$$

$$x = 30 \times \cos 64° = 30 \times 0.4384 = 13.2\,\text{mm}$$

Fig. 7.4

$$\tan 40° = \frac{60}{x}$$

$$x = \frac{60}{\tan 40°} = \frac{60}{0.839} = 71.5\,\text{mm}$$

**EXAMPLE 7.2**

Find the angles marked $\theta$ in Figs. 7.5 and 7.6.

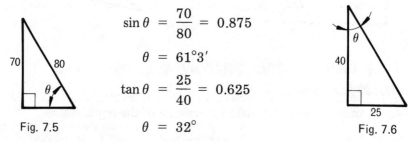

$$\sin \theta = \frac{70}{80} = 0.875$$

$$\theta = 61°3'$$

$$\tan \theta = \frac{25}{40} = 0.625$$

$$\theta = 32°$$

Fig. 7.5                     Fig. 7.6

# Exercise 7.1

Find the lengths of the sides marked $x$ in Fig. 7.7.

1)                     2)                     3)

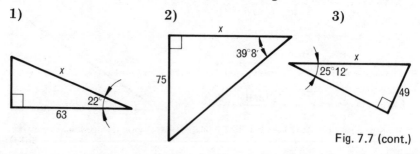

Fig. 7.7 (cont.)

4)                 5)                6)

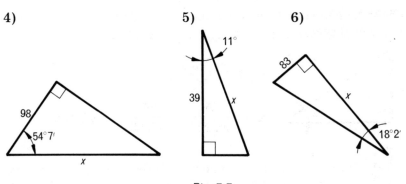

Fig. 7.7

Find the angles marked $\theta$ in Fig. 7.8.

7)          8)                   9)

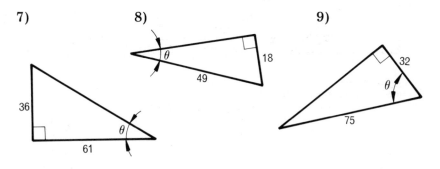

Fig. 7.8

10) The altitude of an isosceles triangle is 86 mm and each of the equal angles is 29°. Calculate the lengths of the equal sides.

# TRIGONOMETRICAL IDENTITIES

A statement of the type $\operatorname{cosec} A \equiv \dfrac{1}{\sin A}$ is called an *identity*.

The sign $\equiv$ means 'is identical to'. Any statement using this sign is true for all values of the variables, i.e. the angle A in the above identity. In practice, however, the $\equiv$ sign is often replaced by the $=$ (equals) sign and the identity would be given as $\operatorname{cosec} A = \dfrac{1}{\sin A}$

Many trigonometrical identities may be verified by the use of a right-angled triangle.

## EXAMPLE 7.3

To show that $\tan A = \dfrac{\sin A}{\cos A}$

The sides and angles of the triangle may be labelled in any way providing that the $90°$ angle is *not* called A. In Fig. 7.9 the standard notation for a triangle has been used.

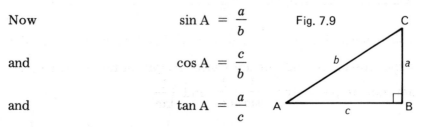

Now $\qquad \sin A = \dfrac{a}{b}$ \qquad Fig. 7.9

and $\qquad \cos A = \dfrac{c}{b}$

and $\qquad \tan A = \dfrac{a}{c}$

Hence from the given identity,

$$\text{RHS} = \frac{\sin A}{\cos A} = \frac{a/b}{c/b} = \frac{ab}{bc} = \frac{a}{c} = \tan A = \text{LHS}$$

## EXAMPLE 7.4

To show that $\sin^2 A + \cos^2 A = 1$

In Fig. 7.9 $\sin A = \dfrac{a}{b}$ \qquad $\therefore$ \quad $\sin^2 A = \left(\dfrac{a}{b}\right)^2 = \dfrac{a^2}{b^2}$

$\cos A = \dfrac{c}{b}$ \qquad $\therefore$ \quad $\cos^2 A = \left(\dfrac{c}{b}\right)^2 = \dfrac{c^2}{b^2}$

$\therefore \qquad \text{LHS} = \sin^2 A + \cos^2 A = \dfrac{a^2}{b^2} + \dfrac{c^2}{b^2} = \dfrac{a^2 + c^2}{b^2}$

But by Pythagoras' theorem, $\quad a^2 + c^2 = b^2$

$\therefore \qquad \text{LHS} = \dfrac{b^2}{b^2} = 1 = \text{RHS}$

Thus $\qquad \sin^2 A + \cos^2 A = 1$

# RECIPROCAL RATIOS

In addition to sin, cos and tan there are three other ratios that may be obtained from a right-angled triangle. These are:

cosecant (called cosec for short)
secant (called sec for short)
cotangent (called cot for short)

The three ratios are defined as follows:

$$\cosec A = \frac{1}{\sin A}$$ $$\sec A = \frac{1}{\cos A}$$ $$\cot A = \frac{1}{\tan A}$$

The reciprocal of $x$ is $\frac{1}{x}$ and it may therefore be seen why the terms cosec, sec and cot are called 'reciprocal ratios', since they are equal respectively to $\frac{1}{\sin}$, $\frac{1}{\cos}$ and $\frac{1}{\tan}$

Formulae in technical reference books often include reciprocal ratios. It will then be necessary for you to re-write the formula before use in terms of the more familiar ratios, namely sin, cos, and tan.

For example, the formula

$$M = D - \frac{5p}{6}\cot\theta + d(\cosec\theta + 1)$$

should be re-written as

$$M = D - \frac{5p}{6}\left(\frac{1}{\tan\theta}\right) + d\left(\frac{1}{\sin\theta} + 1\right)$$

# TRIGONOMETRICAL RATIOS BETWEEN 0° AND 360°

The sine, cosine and tangent of an angle between 0° and 90° have been previously defined. We now show how to deal with angles between 0° and 360°.

From Fig. 7.10 by definition, $\sin\theta = \dfrac{PM}{OP}$

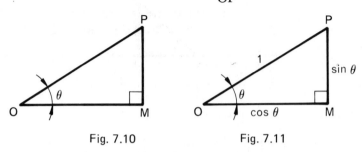

Fig. 7.10                    Fig. 7.11

If we make OP = 1 unit as shown in Fig. 7.11 then $\sin\theta$ = PM and $\cos\theta$ = OM.

In Fig. 7.12 the axes XOX' and YOY' have been drawn at right-angles to each other to form four quadrants shown in the diagram. Drawing the axes in this way allows us to use the same sign convention as when drawing a graph.

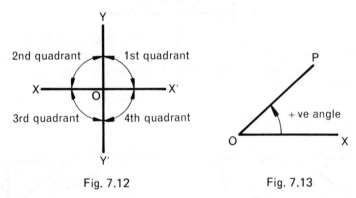

Fig. 7.12                    Fig. 7.13

An angle, if positive, is measured in an anti-clockwise direction from the axis OX, which is the datum line from which all angles are measured. It is formed by rotating a line, such as OP (Fig. 7.13) in an anti-clockwise direction.

Now if we draw a circle whose radius is OP we see that OP, as it rotates, forms the angle $\theta$. If OP is made equal to 1 unit, then by drawing the right-angled triangle OPM (Fig. 7.14) the vertical height PM gives the value of $\sin\theta$ and the horizontal distance OM gives the value of $\cos\theta$.

The idea can be extended to angles greater than $90°$ as shown in Fig. 7.15.

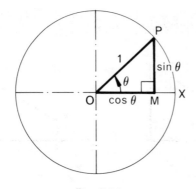

Fig. 7.14

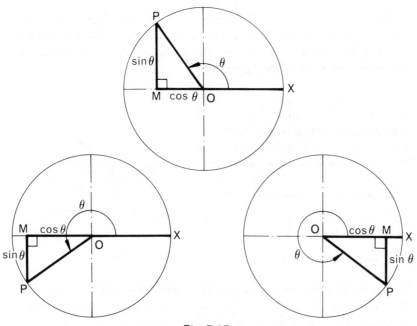

Fig. 7.15

We now make use of sign convention used when drawing a graph. This means that when the height PM lies above the horizontal axis it is a positive length and when it lies below the horizontal axis it is a negative length. Hence when PM lies above the axis XOX' $\sin\theta$ is positive and when it lies below the axis XOX' $\sin\theta$ is negative.

Similarly, when the distance OM lies to the right of the origin O, it is regarded as being a positive distance; if it lies to the left of O it is regarded as being a negative distance. Hence when OM lies to the right of O, $\cos \theta$ is positive and when it lies to the left of O, $\cos \theta$ is negative.

This gives us the signs of the ratios sine and cosine in the four quadrants, see Fig. 7.16.

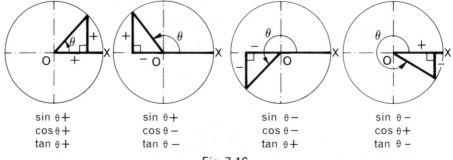

sin θ+      sin θ+      sin θ−      sin θ−
cos θ+      cos θ−      cos θ−      cos θ+
tan θ+      tan θ−      tan θ+      tan θ−

Fig. 7.16

We have shown previously that $\tan \theta \equiv \dfrac{\sin \theta}{\cos \theta}$

Remembering that like signs, when divided, give a positive result and unlike signs give a negative result, we find that $\tan \theta$ is positive in the first and third quadrants and negative in the second and fourth quadrants.

All of the above results are summarised in Fig. 7.17.

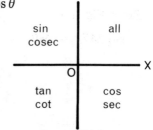

Positive trig. ratios

Fig. 7.17

### EXAMPLE 7.5

Find the values of $\sin 158°$, $\cos 158°$ and $\tan 158°$

As shown in Fig. 7.18, $\sin 158°$ is given by the length PM. But from $\triangle$OPM, PM gives the sine of the angle POM,

$\therefore \quad \sin 158° = \sin \hat{POM}$

$= \sin (180° - 158°)$

$= \sin 22°$

$= 0.3746$

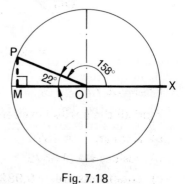

Fig. 7.18

Also from Fig. 7.18, OM gives the value of $\cos 158°$: but this is a negative length since it lies to the left of origin O,

$\therefore$ 
$$\cos 158° = -\cos P\hat{O}M$$
$$= -\cos(180° - 158°)$$
$$= -\cos 22°$$
$$= -0.9272$$

and 
$$\tan 158° = -\tan P\hat{O}M$$
$$= -\tan(180° - 158°)$$
$$= -\tan 22°$$
$$= -0.4040$$

**EXAMPLE 7.6**

Find the values of $\sin 247°$, $\cos 247°$ and $\tan 247°$.

From Fig. 7.19

$$\sin 247° = MP = -\sin P\hat{O}M$$
$$= -\sin(247° - 180°)$$
$$= -\sin 67°$$
$$= -0.9205$$

$$\cos 247° = OM = -\cos P\hat{O}M$$
$$= -\cos 67°$$
$$= -0.3907$$

$$\tan 247° = \frac{MP}{OM} = +\tan P\hat{O}M$$

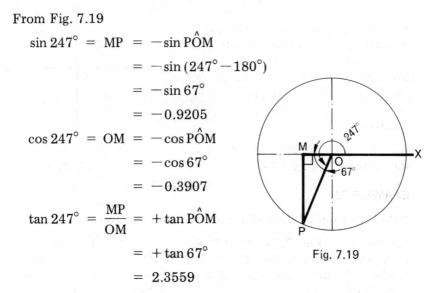

$$= +\tan 67°$$
$$= 2.3559$$

Fig. 7.19

**EXAMPLE 7.7**

Find all the angles between $0°$ and $360°$ whose:

a) sines are $-0.4676$
b) cosines are $0.3572$
c) cotangents are $-0.9827$

**a)** Let $\sin\theta = -0.4676$

Since the sine is negative the angles $\theta$ lie in the 3rd and 4th quadrants. These are shown in Figs. 7.20 and 7.21.

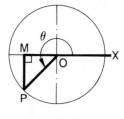
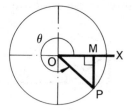

Fig. 7.20                    Fig. 7.21

From right-angled triangle OPM,

$$\sin P\hat{O}M = MP = 0.4676$$

$\therefore$  $$P\hat{O}M = 27°53'\quad\text{from calculator}$$

From Fig. 7.20  $$\theta = 180° + 27°53' = 207°53'$$

and from Fig. 7.21  $$\theta = 360° - 27°53' = 332°7'$$

**b)** Let $\cos\theta = 0.3572$

Since the cosine is positive the angles $\theta$ lie in the 1st and 4th quadrants. These are shown in Figs. 7.22 and 7.23.

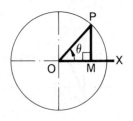

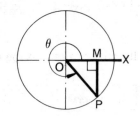

Fig. 7.22                    Fig. 7.23

From the right-angled triangle OPM,

$$\cos P\hat{O}M = OM = 0.3572$$

$\therefore$  $$P\hat{O}M = 69°4'\quad\text{from calculator}$$

From Fig. 7.22  $$\theta = 69°4'$$

and from Fig. 7.23  $$\theta = 360° - 69°4' = 290°56'$$

**c)** Let $\cot\theta = -0.9827$

Cotangents are treated in a similar manner to tangents, i.e. since the cotangent is negative the angles $\theta$ lie in the 2nd and 4th quadrants. These are shown in Figs. 7.24 and 7.25.

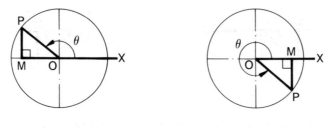

Fig. 7.24                    Fig. 7.25

From the right-angled triangle OPM,

$$\cot P\hat{O}M = \frac{OM}{MP} = 0.9827$$

$\therefore$          $P\hat{O}M = 45°30'$   from calculator

From Fig. 7.24    $\theta = 180° - 45°30' = 134°30'$

and from Fig. 7.25  $\theta = 360° - 45°30' = 314°30'$

The following tables may be used for angles in any quadrant:

| Quadrant | Angle | $\sin\theta =$ |
|---|---|---|
| First | $0°$ to $90°$ | $\sin\theta$ |
| Second | $90°$ to $180°$ | $\sin(180° - \theta)$ |
| Third | $180°$ to $270°$ | $-\sin(\theta - 180°)$ |
| Fourth | $270°$ to $360°$ | $-\sin(360° - \theta)$ |
| Quadrant | $\cos\theta =$ | $\tan\theta =$ |
| First | $\cos\theta$ | $\tan\theta$ |
| Second | $-\cos(180° - \theta)$ | $-\tan(180° - \theta)$ |
| Third | $-\cos(\theta - 180°)$ | $\tan(\theta - 180°)$ |
| Fourth | $\cos(360° - \theta)$ | $-\tan(360° - \theta)$ |

## Exercise 7.2

1) Write down the values of the sine, cosine and tangent of the following angles:

(a) $121°$     (b) $178°23'$   (c) $102°29'$   (d) $211°$

(e) $239°17'$   (f) $258°28'$   (g) $318°27'$   (h) $297°17'$

2) Evaluate: $6 \sin 23° - 2 \cos 47° + 3 \tan 17°$.

3) Evaluate: $5 \sin 142° - 3 \tan 148° + 3 \cos 230°$.

4) Evaluate: $\sin A \cos B - \sin B \cos A$
given that $\sin A = \frac{3}{5}$ and $\tan B = \frac{4}{3}$. A and B are both acute angles.

5) An angle A is in the 2nd quadrant. If $\sin A = \frac{3}{5}$ find, without actually finding angle A, the value of $\cos A$ and $\tan A$.

6) If $\sin \theta = 0.1432$ find all the values of $\theta$ from $0°$ to $360°$.

7) If $\cos \theta = -0.8927$ find all the values of $\theta$ from $0°$ to $360°$.

8) Find the angles in the first and second quadrants:

(a) whose sine is $0.7137$          (b) whose cosine is $-0.4813$

(c) whose tangent is $0.9476$       (d) whose tangent is $-1.7642$

9) Find the angles in the third or fourth quadrants:

(a) whose sine is $-0.7880$         (b) whose cosine is $0.5592$

(c) whose tangent is $-2.9042$

10) If $\sin A = \dfrac{a \sin B}{b}$ find the values of A between $0°$ and $360°$
when $a = 7.26 \, \text{mm}$, $b = 9.15 \, \text{mm}$ and $B = 18°29'$

**11)** If $\cos C = \dfrac{(a^2 + b^2 - c^2)}{2ab}$ find the values of C between $0°$ and $360°$ given that $a = 1.26\,\text{m}$, $b = 1.41\,\text{m}$ and $c = 2.13\,\text{m}$.

## SINE, COSINE AND TANGENT CURVES

A sine curve (or sine waveform) is the result of plotting vertically the values of $\sin\theta$ against $\theta$ horizontally (often called an angle base). The values of $\sin\theta$ may be found using a scientific calculator, but an alternative method is shown in Fig. 7.26. This makes use of the ideas expressed in Figs 7.14 and 7.15.

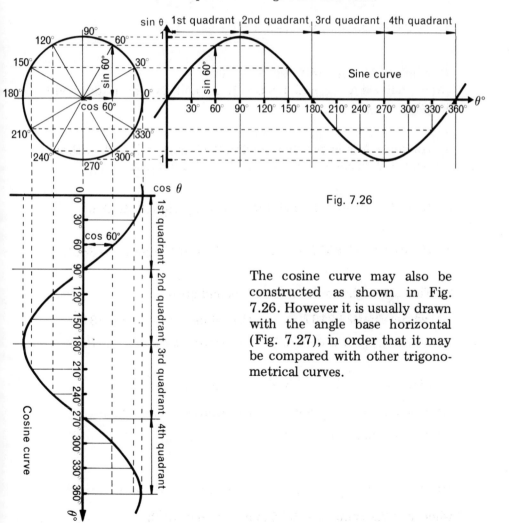

Fig. 7.26

The cosine curve may also be constructed as shown in Fig. 7.26. However it is usually drawn with the angle base horizontal (Fig. 7.27), in order that it may be compared with other trigonometrical curves.

## The Sine Curve

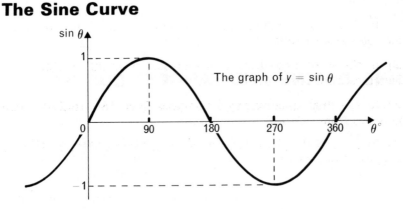

Fig. 7.27

The following features should be noted:

**(1)** In the first quadrant as $\theta$ increases from $0°$ to $90°$, $\sin\theta$ increases from 0 to 1

**(2)** In the second quadrant as $\theta$ increases from $90°$ to $180°$, $\sin\theta$ decreases from 1 to 0

**(3)** In the third quadrant as $\theta$ increases from $180°$ to $270°$, $\sin\theta$ decreases from 0 to $-1$

**(4)** In the fourth quadrant as $\theta$ increases from $270°$ to $360°$, $\sin\theta$ increases from $-1$ to 0

## The Cosine Curve

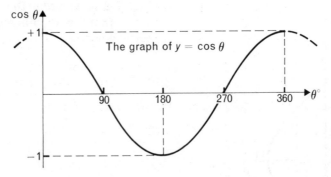

Fig. 7.28

Note that:

**(1)** In the first quadrant as $\theta$ increases from $0°$ to $90°$, $\cos \theta$ decreases from 1 to 0

**(2)** In the second quadrant as $\theta$ increases from $90°$ to $180°$, $\cos \theta$ decreases from 0 to $-1$

**(3)** In the third quadrant as $\theta$ increases from $180°$ to $270°$, $\cos \theta$ increases from $-1$ to 0

**(4)** In the fourth quadrant as $\theta$ increases from $270°$ to $360°$, $\cos \theta$ increases from 0 to 1

## The Tangent Curve

The graph of $y = \tan \theta$ may be drawn using values obtained from the identity $\tan \theta = \dfrac{\sin \theta}{\cos \theta}$. A table of values may be drawn up part of which is given below.

You should remember that:

$$\frac{1}{\text{Very small number}} = \text{Very large number}$$

Thus:

$$\frac{1}{\text{Zero}} = \text{Infinity (symbol } \infty\text{)}$$

| $\theta°$ | 0 | 10 | 20 | 30 | 40 |
|---|---|---|---|---|---|
| $\sin \theta$ | 0 | 0.174 | 0.342 | 0.500 | 0.643 |
| $\cos \theta$ | 1 | 0.985 | 0.940 | 0.866 | 0.766 |
| $\tan \theta = \dfrac{\sin \theta}{\cos \theta}$ | 0 | 0.176 | 0.364 | 0.577 | 0.839 |

| $\theta°$ | 50 | 60 | 70 | 80 | 90 |
|---|---|---|---|---|---|
| $\sin \theta$ | 0.766 | 0.866 | 0.940 | 0.985 | 1 |
| $\cos \theta$ | 0.643 | 0.500 | 0.342 | 0.174 | 0 |
| $\tan \theta = \dfrac{\sin \theta}{\cos \theta}$ | 1.19 | 1.73 | 2.75 | 5.67 | $\infty$ |

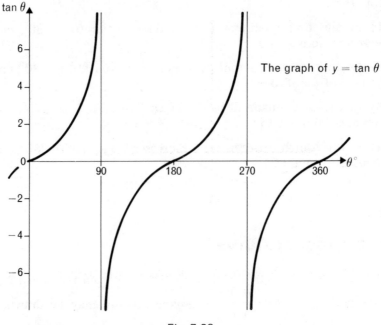

Fig. 7.29

Note that:

**(1)** In the first quadrant as $\theta$ increases from $0°$ to $90°$, $\tan \theta$ increases from 0 to infinity

**(2)** In the second quadrant as $\theta$ increases from $90°$ to $180°$, $\tan \theta$ increases from minus infinity to 0

**(3)** In the third quadrant as $\theta$ increases from $180°$ to $270°$, $\tan \theta$ increases from 0 to infinity

**(4)** In the fourth quadrant as $\theta$ increases from $270°$ to $360°$, $\tan \theta$ increases from minus infinity to 0

# Exercise 7.3

**1)** Draw the graphs of (i) $y = \sin \theta$, (ii) $y = \cos \theta$ for values of $\theta$ between $0°$ and $360°$. From the graphs find values of the sine and cosine of the angles:

(a) $38°$      (b) $72°$      (c) $142°$      (d) $108°$

(e) $200°$     (f) $250°$     (g) $305°$      (h) $328°$

**2)** Plot the graph of $y = 3 \sin x$ between $0°$ and $360°$. From the graph read off the values of $x$ for which $y = 1.50$ and find the value of $y$ when $x = 250°$.

**3)** Draw the graphs of $3 \cos \theta$ for values of $\theta$ from $0°$ to $360°$. Use the graph to find approximate values of the two angles for which $3 \cos \theta = 0.6$

**4)** By projection from the circumference of a suitably marked off circle, draw the graph of $4 \sin \theta$ for values of $\theta$ from $0°$ to $360°$. Use the graph to find approximate values of the two angles for which $4 \sin \theta = 1.6$

## THE SOLUTION OF TRIANGLES

We now deal with triangles which are *not* right-angled. Every triangle consists of six elements—three angles and three sides.

If we are given any three of these six elements we can find the other three by using either the *Sine Rule* or the *Cosine Rule*. (The exception is when we are given three angles, since it is obvious that a triangle of a given shape can be of any size.)

When we have found the values of the three missing elements we are said to have 'solved the triangle'.

## THE SINE RULE

The sine rule may be used when given:

**(1)** one side and any two angles; or

**(2)** two sides and an angle opposite to one of the given sides. (In this case two solutions may be found giving rise to what is called the 'ambiguous case', see Example 7.9.)

Using the notation of Fig. 7.30 the *sine rule* states:

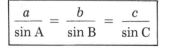

$$\frac{a}{\sin A} = \frac{b}{\sin B} = \frac{c}{\sin C}$$

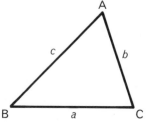

Fig. 7.30

**EXAMPLE 7.8**

Solve the triangle ABC given that
A = 42°, C = 72° and $b = 61.8$ mm.

The triangle should be drawn for
reference as shown in Fig. 7.31,
but there is no need to draw it to
scale.

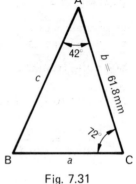

Since $\angle A + \angle B + \angle C = 180°$

$\angle B = 180° - 42° - 72° = 66°$

Fig. 7.31

The sine rules states:

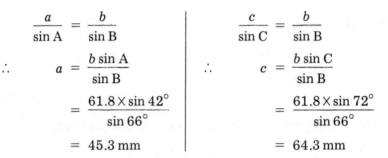

$$\frac{a}{\sin A} = \frac{b}{\sin B} \qquad\qquad \frac{c}{\sin C} = \frac{b}{\sin B}$$

$$\therefore \quad a = \frac{b \sin A}{\sin B} \qquad\qquad \therefore \quad c = \frac{b \sin C}{\sin B}$$

$$= \frac{61.8 \times \sin 42°}{\sin 66°} \qquad\qquad = \frac{61.8 \times \sin 72°}{\sin 66°}$$

$$= 45.3 \text{ mm} \qquad\qquad = 64.3 \text{ mm}$$

The complete solution is:

$$\angle B = 66°, \quad a = 45.3 \text{ mm}, \quad c = 64.3 \text{ mm}$$

A rough check on sine rule calculations may be made by
remembering that in any triangle the longest side lies opposite
the largest angle and the shortest side lies opposite the smallest
angle.

Thus in the previous example:

Smallest angle $= 42° = $ A;     Shortest side $= a = 45.3$ mm

Largest angle $= 72° = $ C;     Longest side $= c = 64.3$ mm

## The Ambiguous Case

There are two angles between 0° and 180° which have the same
sine. For instance if sin A = 0.5000, then A can be either 30° or
150°. When using the sine rule to find an angle we must always
examine the problem to see if there are two possible values for the
angle.

**EXAMPLE 7.9**

In triangle ABC, $b = 93.23$ mm,
$c = 85.61$ mm and $\angle C = 37°$.

Solve the triangle.

Referring to Fig. 7.32 we have

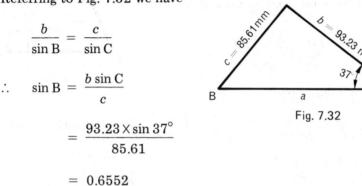

$$\frac{b}{\sin B} = \frac{c}{\sin C}$$

$\therefore \quad \sin B = \dfrac{b \sin C}{c}$

$$= \frac{93.23 \times \sin 37°}{85.61}$$

$$= 0.6552$$

Fig. 7.32

The angle B may be in either the first or second quadrants.

In the first quadrant, $\angle B = 40°56'$   (Fig. 7.33).

In the second quadrant, $\angle B = 139°4'$   (Fig. 7.34).

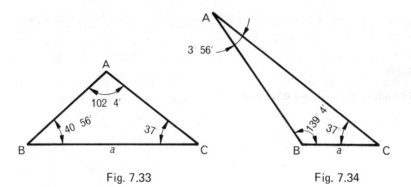

Fig. 7.33                    Fig. 7.34

When

$\angle B = 40°56'$

$\angle A = 180° - 40°56' - 37°$

$\quad = 102°4'$

When

$\angle B = 139°4'$

$\angle A = 180° - 139°4' - 37°$

$\quad = 3°56'$

Now

$$\frac{a}{\sin A} = \frac{c}{\sin C}$$

$$\therefore \quad a = \frac{c \sin A}{\sin C}$$

$$= \frac{85.61 \sin 102°4'}{\sin 37°}$$

$$= \frac{85.61 \sin 77°56'}{\sin 37°}$$

$$= 130 \, \text{mm}$$

Now

$$\frac{a}{\sin A} = \frac{c}{\sin C}$$

$$\therefore \quad a = \frac{c \sin A}{\sin C}$$

$$= \frac{85.61 \sin 3°56'}{\sin 37°}$$

$$= 9.77 \, \text{mm}$$

The ambiguous case may be seen clearly by constructing the given triangle geometrically as follows (Fig. 7.35).

Using a full size scale draw AC = 93.23 mm and draw CX such that ACX = 37°. Now with centre A and radius 85.61 mm describe a circular arc to cut CX at B and B'.

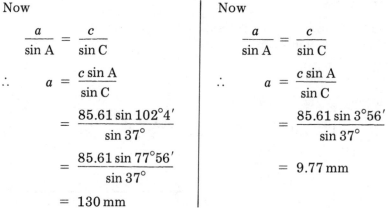

Fig. 7.35

Then ABC represents the triangle shown in Fig. 7.33 and AB'C represents the triangle shown in Fig. 7.34.

## Use of the Sine Rule to Find the Diameter (D) of the Circumscribing Circle of a Triangle

Using the notation of Fig. 7.36.

$$\frac{a}{\sin A} = \frac{b}{\sin B} = \frac{c}{\sin C} = D$$

The rule is useful when we wish to find the pitch circle diameter of a ring of holes.

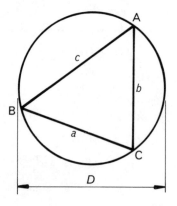

Fig. 7.36

**EXAMPLE 7.10**

In Fig. 7.37 three holes are positioned by the angle and dimensions shown. Find the pitch circle diameter.

We are given

$$\angle B = 41° \quad \text{and} \quad b = 112.5 \text{ mm}$$

$$\therefore \quad D = \frac{b}{\sin B} = \frac{112.5}{\sin 41°}$$

$$= 171.5 \text{ mm}$$

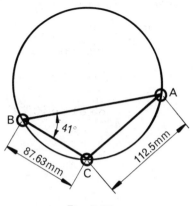

Fig. 7.37

# THE COSINE RULE

The cosine rule is used in all cases where the sine rule cannot be used. These are when given:

**(1)** two sides and the angle between them;

**(2)** three sides.

Whenever possible the sine rule is used because it results in a calculation which is easier to perform. In solving a triangle it is sometimes necessary to start with the cosine rule and then having found one of the unknown elements to finish solving the triangle using the sine rule.

The cosine rules states:

either

or

or

$$a^2 = b^2 + c^2 - 2bc \cos A$$
$$b^2 = a^2 + c^2 - 2ac \cos B$$
$$c^2 = a^2 + b^2 - 2ab \cos C$$

**EXAMPLE 7.11**

Solve the triangle ABC if $a = 70$ mm, $b = 40$ mm and $\angle C = 64°$.

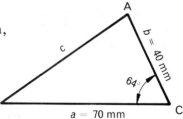

Fig. 7.38

Referring to Fig. 7.38, to find the side $c$ we use

$$c^2 = a^2 + b^2 - 2ab \cos C$$
$$= 70^2 + 40^2 - 2 \times 70 \times 40 \times \cos 64° = 4044$$

$\therefore$ $\qquad c = \sqrt{4044} = 63.6 \, \text{mm}$

We now use the sine rule to find $\angle A$:

$$\frac{a}{\sin A} = \frac{c}{\sin C}$$

$\therefore$ $\qquad \sin A = \dfrac{a \sin C}{c} = \dfrac{70 \times \sin 64°}{63.6}$

Thus $\qquad A = 81°36'$

and $\qquad B = 180° - 81°36' - 64° = 34°24'$

### EXAMPLE 7.12

The mast AB of a job crane (Fig. 7.39) is 3 m long and the tie BC is 2.4 m long. The angle between AB and BC is 125°. Find the length of the jib AC.

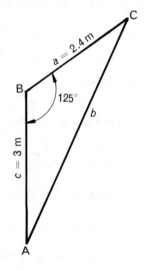

Fig. 7.39

Using the cosine rule:

$$b^2 = a^2 + c^2 - 2ac \cos B$$
$$= 2.4^2 + 3^2 - 2 \times 2.4 \times 3 \times \cos 125°$$

Now $\quad \cos 125° = -\cos(180° - 125°) = -\cos 55° = -0.5736$

Hence $\qquad b^2 = 2.4^2 + 3^2 - 2 \times 2.4 \times 3 \times (-0.5736)$

$\therefore$ $\qquad b = 4.80$

Therefore the jib of the crane is 4.80 m long.

**EXAMPLE 7.13**

The instantaneous values, $i_1$ and $i_2$, of two alternating currents are represented by the two sides of a triangle shown in Fig. 7.40. The resultant current $i_R$ is represented by the third side. Calculate the magnitude of $i_R$ and the angle $\phi$ between the current $i_1$ and $i_R$.

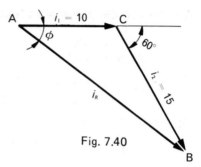

In $\triangle ABC$, Fig. 7.40, we have

$b = 10$, $a = 15$ and $\angle C = 120°$.      Fig. 7.40

Using the cosine rule gives

$$c^2 = a^2 + b^2 - 2ab \cos C$$
$$= 15^2 + 10^2 - 2 \times 15 \times 10 \times \cos 120°$$
$$= 225 + 100 - 300 \times (-\cos 60°)$$
$$\therefore \qquad c = \sqrt{475} = 21.79 = i_R$$

To find $\angle A$ we use the sine rule,

$$\frac{a}{\sin A} = \frac{c}{\sin C}$$

$$\therefore \quad \sin A = \frac{a \sin C}{c} = \frac{15 \times \sin 120°}{21.79} = \frac{15 \times \sin 60°}{21.79} = 0.5962$$

$$\therefore \qquad A = 39°56' = \phi$$

Hence the magnitude of $i_R$ is 21.8 and the angle $\phi$ is 36°36′.

# Exercise 7.4

1) The following are all exercises on the sine rule. Solve the following triangles ABC given:

(a)  $A = 75°$          $B = 34°$              $a = 102$ mm
(b)  $C = 61°$          $B = 71°$              $b = 91$ mm
(c)  $A = 19°$          $C = 105°$            $c = 11.1$ m
(d)  $A = 116°$        $C = 18°$              $a = 170$ mm

| (e) | $A = 36°$ | $B = 77°$ | $b = 2.5\,\text{m}$ |
| (f) | $A = 49°11'$ | $B = 67°17'$ | $c = 11.22\,\text{mm}$ |
| (g) | $A = 17°15'$ | $C = 27°7'$ | $b = 221.5\,\text{mm}$ |
| (h) | $A = 77°3'$ | $C = 21°3'$ | $a = 9.793\,\text{m}$ |
| (i) | $B = 115°4'$ | $C = 11°17'$ | $c = 516.2\,\text{mm}$ |
| (j) | $a = 17\,\text{m}$ | $b = 15\,\text{m}$ | $B = 39°$ |
| (k) | $a = 7\,\text{m}$ | $c = 11\,\text{m}$ | $C = 22°7'$ |
| (l) | $b = 92\,\text{mm}$ | $c = 71\,\text{mm}$ | $C = 39°8'$ |
| (m) | $b = 15.13\,\text{m}$ | $c = 11.62\,\text{m}$ | $B = 85°17'$ |
| (n) | $a = 23\,\text{m}$ | $c = 18.2\,\text{m}$ | $A = 49°19'$ |
| (o) | $a = 9.217\,\text{m}$ | $b = 7.152\,\text{m}$ | $A = 105°4'$ |

**2)** Solve the following triangles ABC using the cosine rule:

| (a) | $a = 9\,\text{m}$ | $b = 11\,\text{m}$ | $C = 60°$ |
| (b) | $b = 10\,\text{m}$ | $c = 14\,\text{m}$ | $A = 56°$ |
| (c) | $a = 8.16\,\text{m}$ | $c = 7.14\,\text{m}$ | $B = 37°18'$ |
| (d) | $a = 5\,\text{m}$ | $b = 8\,\text{m}$ | $c = 7\,\text{m}$ |
| (e) | $a = 312\,\text{mm}$ | $b = 527.3\,\text{mm}$ | $c = 700\,\text{mm}$ |
| (f) | $a = 7.912\,\text{m}$ | $b = 4.318\,\text{m}$ | $c = 11.08\,\text{m}$ |

**3)** Three holes lie on a pitch circle and their chordal distances are 41.82 mm, 61.37 mm and 58.29 mm. Find their pitch circle diameter.

**4)** In Fig. 7.41 find the angle BCA given that BC is parallel to AD.

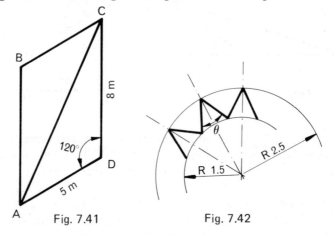

Fig. 7.41                Fig. 7.42

**5)** Calculate the angle $\theta$ in Fig. 7.42. There are 12 castellations and they are equally spaced.

**6)** Find the smallest angle in a triangle whose sides are 20, 25 and 30 m long.

**7)** In Fig. 7.43 find:

(a) the distance AB                    (b) the angle ACB

**8)** Three holes are spaced in a plate detail as shown in Fig. 7.44. Calculate the centre distances from A to B and from A to C.

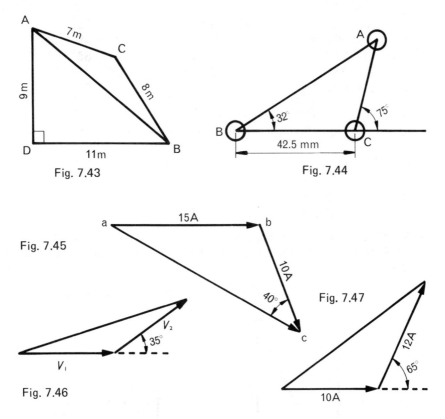

Fig. 7.43

Fig. 7.44

Fig. 7.45

Fig. 7.47

Fig. 7.46

**9)** In Fig. 7.45, *ab* and *bc* are phasors representing the alternating currents in two branches of a circuit. The line *ac* represents the resultant current. Find by calculation this resultant current.

**10)** Two phasors are shown in Fig. 7.46. If $V_1 = 8$ and $V_2 = 6$ calculate the value of their resultant and the angle it makes with $V_1$.

**11)** Calculate the resultant of the two phasors shown in Fig. 7.47.

**12)** The main chain lines of a small survey are shown in Fig. 7.48. Calculate the angle $\theta$.

**13)** A part of a field survey consists of a triangle ABC in which the sides AB, BC and CA measure 36 m, 40 m and 47 m respectively. Find the size of each angle in the triangle.

**14)** Fig. 7.49 represents part of a roof truss. Calculate the length of the member BC.

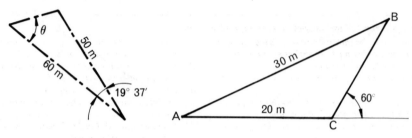

Fig. 7.48                    Fig. 7.49

# STATISTICS

*On reaching the end of this chapter you should be able to:*

1. *Define the arithmetic mean, median and mode, and explain where each is an appropriate measure of central tendency.*
2. *Calculate the arithmetic mean for ungrouped data.*
3. *Place ungrouped data in rank order and determine the median and modal values.*
4. *Calculate the arithmetic mean for grouped data.*
5. *Estimate the mode of grouped data using a histogram.*

6. *Determine the median, quartiles and percentiles from cumulative frequency data.*
7. *Describe the concepts of variability of data and the need to measure scatter.*
8. *Define the standard deviation.*
9. *Calculate the standard deviation for ungrouped data.*
10. *Calculate the standard deviation for grouped data with equal intervals.*

## VARIABLES

Variables are measured quantities which can be expressed as numbers. For instance, the diameter of a workpiece or the resistance of an electrical component.

Some variables can only take certain values. For instance, the number of people working in a factory can only be a whole number, say 235. It cannot be 235.4, etc. Such variables are said to be *discrete* or *discontinuous*. Some other examples are the number of £1 notes in the wages office of a factory or the sizes of footwear. Note that the values of a discrete variable need not necessarily be a whole number. The sizes of shoes usually vary by half-sizes so that shoes size 5, $5\frac{1}{2}$, 6, $6\frac{1}{2}$, 8, etc. might be on sale. However, shoes size 5.24 will not be available and hence the size of shoes is a discrete variable.

Other variables can take up any value between two given end values. Suppose that a resistance has to be $125 \pm 0.2$ ohms. The resistance may have any value between the end values of 124.8 ohms and 125.2 ohms. Variables, such as the value of a resistance, are said to be *continuous* variables.

In statistics a variable is sometimes called the *variate*, particularly when dealing with frequency distributions and histograms.

## VARIABILITY

Variations in size always occur when articles are manufactured. Because of the variability of every manufacturing method the articles produced by a single process will almost certainly differ from each other.

It is not possible, or desirable, to measure manufactured parts to extreme accuracy and measurements which are made are accurate only to the limits imposed by the measuring device. For instance most micrometer readings are made to an accuracy of 0.01 mm.

## FREQUENCY DISTRIBUTIONS

Consider a bottling plant at a pharmaceutical firm dispensing a nominal 50 mℓ mixture. From a sample batch of 90 bottles the results might be:

| | | | | | | | |
|---|---|---|---|---|---|---|---|
| 50.07 | 49.92 | 50.09 | 50.00 | 50.05 | 50.02 | 50.05 | 50.02 |
| 49.94 | 49.95 | 49.99 | 50.04 | 50.04 | 50.02 | 50.00 | 49.95 |
| 50.08 | 49.98 | 49.98 | 50.01 | 50.01 | 49.96 | 50.06 | 50.04 |
| 49.97 | 50.00 | 50.07 | 50.05 | 49.99 | 50.06 | 50.03 | 50.03 |
| 49.96 | 50.05 | 50.02 | 50.01 | 49.99 | 50.06 | 50.04 | 50.02 |
| 50.01 | 50.02 | 50.02 | 49.96 | 49.99 | 50.01 | 50.03 | 50.03 |
| 50.02 | 50.00 | 49.97 | 49.97 | 50.00 | 50.05 | 50.01 | 50.06 |
| 49.94 | 49.99 | 50.03 | 49.98 | 49.99 | 50.03 | 50.04 | 50.01 |
| 50.01 | 50.01 | 49.93 | 50.02 | 50.05 | 50.08 | 49.98 | 49.95 |
| 50.03 | 50.04 | 49.97 | 50.00 | 50.04 | 50.01 | 50.07 | 50.09 |
| 49.93 | 50.00 | 50.02 | 50.01 | 49.98 | 49.96 | 49.98 | 50.03 |
| 49.97 | 50.00 | | | | | | |

The figures do not mean very much as they stand and so we must arrange them into an ordered form, i.e. into a frequency distribution. To do this we collect together all the 49.92 mℓ readings, all the 49.93 mℓ readings etc. A tally chart (Table 1) is the best way of doing this. Each time a measurement arises a tally mark is placed opposite the appropriate measurement. The fifth tally mark is usually made in an oblique direction thus tying the tally marks into bundles of five, which makes for easier counting. When the tally marks are complete, the marks are counted and the numerical value recorded in the column headed 'frequency', the frequency being the number of times each measurement occurs.

From Table 1 we see that 49.92 mℓ occurs once (i.e. a frequency of 1), 50.04 mℓ occurs seven times (i.e. a frequency of 7) etc.

TABLE 1

| Mixture (mℓ) | Number of bottles of this size | Frequency |
|---|---|---|
| 49.92 | \| | 1 |
| 49.93 | \| \| | 2 |
| 49.94 | \| \| | 2 |
| 49.95 | \| \| \| | 3 |
| 49.96 | \| \| \| \| | 4 |
| 49.97 | ⊔⊔⊓ | 5 |
| 49.98 | ⊔⊔⊓  \| | 6 |
| 49.99 | ⊔⊔⊓  \| | 6 |
| 50.00 | ⊔⊔⊓  \| \| \| | 8 |
| 50.01 | ⊔⊔⊓  ⊔⊔⊓  \| | 11 |
| 50.02 | ⊔⊔⊓  ⊔⊔⊓ | 10 |
| 50.03 | ⊔⊔⊓  \| \| \| | 8 |
| 50.04 | ⊔⊔⊓  \| \| | 7 |
| 50.05 | ⊔⊔⊓  \| | 6 |
| 50.06 | \| \| \| \| | 4 |
| 50.07 | \| \| \| | 3 |
| 50.08 | \| \| | 2 |
| 50.09 | \| \| | 2 |

# THE HISTOGRAM

The frequency distribution becomes even more understandable if we draw a diagram to represent it. The best type of diagram is the histogram (Fig. 8.1) which consists of a set of rectangles each of the same width, whose heights represent the frequencies. On studying the diagram the pattern of the variation is easily understood, most of the values being grouped near the centre of the diagram with a few values more widely dispersed.

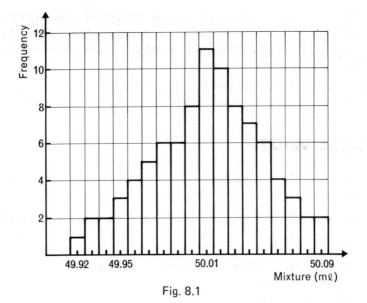

Fig. 8.1

## GROUPED DATA

When dealing with a large amount of data it is often useful to group the information into classes or categories. Referring to Table 1, we could re-arrange the data into a grouped frequency distribution as follows.

TABLE 2

| Class | Frequency |
|-------|-----------|
| 49.92–49.94 | 5 |
| 49.95–49.97 | 12 |
| 49.98–50.00 | 20 |
| 50.01–50.03 | 29 |
| 50.04–50.06 | 17 |
| 50.07–50.09 | 7 |

The first class consists of bottle mixtures between 49.92 mℓ and 49.94 mℓ. Since five bottles belong to this class, the class frequency is 5. Similarly the other class frequencies may be calculated.

## CLASS BOUNDARIES

If we had measured the volumes of mixtures in the bottles of Table 2 to an accuracy of three decimal places then in the first class we would have put all the volumes between 49.915 mℓ and

49.945 mℓ. These values are called the lower and upper class boundaries respectively. Similarly for example, for the fourth class the lower class boundary is 50.005 and the upper class boundary is 50.035.

## CLASS WIDTH

The width of the class is the difference between the lower and upper class boundaries, that is

Class width = Upper class boundary − Lower class boundary

For the first class:

Class width = $49.945 - 49.915 = 0.03$ mℓ

## HISTOGRAM FOR A GROUPED DISTRIBUTION

A histogram for a grouped distribution may be drawn by using the lower and upper class boundaries as the extreme values for each rectangle making up the diagram. Strictly, the areas of the rectangles represent the frequencies, but if the class widths are all the same it is usual to make the heights of the rectangles equal to the class frequencies. Fig. 8.2 shows the histogram drawn for the grouped distribution of Table 2.

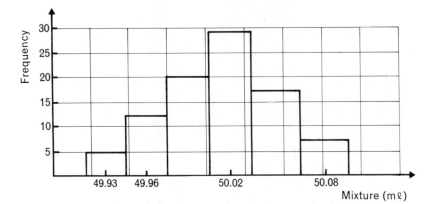

Fig. 8.2

## Exercise 8.1

1) The following figures are the hottest daily temperatures (°C) during June at a particular coastal resort:

| 20, 21, 19, 22, 22, 23, 23, 23, 24, 25, 25, 26, 27, 28, 25 |
| 24, 24, 23, 22, 21, 22, 23, 23, 24, 25, 24, 25, 26, 27, 26 |

With the aid of a frequency table draw a histogram for these temperatures.

2) During trials of one variety of broad bean plant a seed merchant noted the number of beans in each pod as listed below:

| 7, 10, 6, 7, 10, 7, 8, 9, 7, 7, 6, 5, 7, 7, 3, |
| 6, 9, 5, 11, 4, 7, 5, 4, 10, 7, 7, 6, 8, 5, 6, |
| 9, 9, 8, 8, 7, 4, 8, 6, 5, 8, 7, 7, 4, 6, 2, |
| 8, 5, 7, 9, 5, 5, 8, 5, 6, 6, 6, 8, 8, 9, 5, |
| 8, 6, 6, 9, 9, 6, 9, 8, 9, 6, 7, 6, 8, 6, 8, |
| 6, 8, 6, 8, 7, 7, 7, 7, 7, 9, 7, 6, 7, 6, 7, |
| 7, 10, 10, 7, 7, 9, 6, 8, 11, 8 |

Draw up a frequency table and hence construct a histogram for the number of beans/pod.

3) Group the distribution in Question 2 into 5 classes. Hence construct an amended frequency table and histogram. Calculate the class width.

4) For the grouped frequency distribution given below, draw a histogram and state the class width for each of the classes.

| Resistance (ohms) | 110–112 | 113–115 | 116–118 | 119–121 | 122–124 |
|---|---|---|---|---|---|
| Frequency | 2 | 8 | 15 | 9 | 3 |

5) A sample of 100 domestic light bulbs are tested for their durability during pre-marketing trials. The following table, as a grouped frequency distribution, shows the results:

| Life (hours) | 200–299 | 300–399 | 400–499 |
|---|---|---|---|
| Frequency | 8 | 18 | 27 |

| Life (hours) | 500–599 | 600–699 | 700–799 |
|---|---|---|---|
| Frequency | 25 | 14 | 8 |

(a) Calculate the class width.
(b) Draw a histogram for this distribution.

**6)** The lengths of 40 common building bricks were measured to the nearest millimetre with the following results:

| | | | | | | |
|---|---|---|---|---|---|---|
| 214 | 216 | 215 | 216 | 217 | 215 | 216 |
| 215 | 213 | 214 | 215 | 215 | 216 | 215 |
| 216 | 215 | 215 | 217 | 214 | 215 | 216 |
| 212 | 214 | 216 | 215 | 216 | 214 | 216 |
| 215 | 217 | 213 | 215 | 215 | 215 | 214 |
| 214 | 215 | 218 | 213 | | | |

Draw up a frequency table and hence draw a histrogram for these measurements.

**7)** 80 houses were checked to find the number of power points per house. The results were as follows:

5, 8, 3, 5, 5, 6, 3, 0, 6, 2, 5, 2, 2, 1, 9, 6, 8, 6, 3, 6,
4, 2, 7, 4, 5, 4, 5, 8, 2, 8, 8, 5, 2, 5, 7, 5, 3, 7, 6, 2,
8, 6, 6, 3, 5, 5, 7, 9, 6, 4, 8, 3, 2, 6, 4, 9, 2, 8, 6, 4,
6, 3, 1, 8, 2, 4, 4, 1, 1, 9, 5, 7, 7, 5, 5, 7, 2, 6, 2, 4

Draw up a frequency table and hence draw a histrogram of this data.

**8)** The table below gives a grouped frequency distribution for the compressive strength of a certain type of load-carrying brick.

| Strength $(N/mm^2)$ | 59.4–59.6 | 59.7–59.9 | 60.0–60.2 |
|---|---|---|---|
| Frequency | 8 | 37 | 90 |

| Strength $(N/mm^2)$ | 60.3–60.5 | 60.6–60.8 |
|---|---|---|
| Frequency | 52 | 13 |

Determine the class width and draw a histrogram for this distribution.

**9)** A patrol inspector visits an automátic lathe once every six minutes. He picks up a component as it drops into the hopper and measures its diameter. After a shift of 4 hours he had collected a sample of 40 components whose diameters are:

| | | | | | | |
|---|---|---|---|---|---|---|
| 24.98 | 24.96 | 24.97 | 24.98 | 24.99 | 24.97 | 25.03 |
| 25.00 | 24.99 | 25.01 | 25.03 | 25.01 | 25.01 | 25.00 |
| 25.02 | 25.02 | 25.00 | 25.02 | 25.01 | 25.04 | 25.02 |
| 25.02 | 25.01 | 24.97 | 24.98 | 25.01 | 25.03 | 24.99 |
| 25.03 | 25.05 | 24.95 | 24.98 | 24.99 | 25.00 | 25.01 |
| 24.99 | 25.02 | 24.97 | 25.04 | 25.00 | | |

Draw up a frequency table and hence draw a histogram for these measurements.

10) The lengths of 80 machined parts are measured with the following results (the measurements are given in hundredths of a millimetre above 62.0 mm).

| |
|---|
| 5, 8, 3, 5, 5, 6, 3, 0, 6, 2, 5, 2, 2, 1, 9, 6, 8, 6, 3, 6, 4, 2, 7, 4, 5, 4, 5, 8, 2, 8, 8, 5, 2, 5, 7, 5, 3, 7, 6, 2, 8, 6, 6, 3, 5, 5, 7, 9, 6, 4, 8, 3, 2, 6, 4, 9, 2, 8, 6, 4, 6, 3, 1, 8, 2, 4, 4, 1, 1, 9, 5, 7, 7, 5, 5, 7, 2, 6, 2, 4 |

Draw up a frequency table and hence draw a histogram of this data.

11) The table below gives a grouped frequency distribution.

| Diameter (mm) | 5.94-5.96 | 5.97-5.99 | 6.00-6.02 |
|---|---|---|---|
| Frequency | 5 | 34 | 87 |

| Diameter (mm) | 6.03-6.05 | 6.06-6.08 |
|---|---|---|
| Frequency | 49 | 10 |

(a) Determine the class width.
(b) Draw a histogram for this distribution.

12) The data below was obtained by measuring the frequencies (in kilohertz) of 60 tuned circuits. Construct a frequency distribution and hence draw a histrogram to represent the distribution.

| | | | | | | | |
|---|---|---|---|---|---|---|---|
| 12.37 | 12.29 | 12.40 | 12.41 | 12.31 | 12.35 | 12.37 | 12.35 |
| 12.33 | 12.36 | 12.32 | 12.36 | 12.40 | 12.38 | 12.33 | 12.35 |
| 12.30 | 12.30 | 12.34 | 12.39 | 12.44 | 12.32 | 12.27 | 12.32 |
| 12.41 | 12.40 | 12.37 | 12.46 | 12.35 | 12.34 | 12.38 | 12.43 |
| 12.36 | 12.35 | 12.26 | 12.28 | 12.36 | 12.24 | 12.42 | 12.39 |
| 12.45 | 12.42 | 12.28 | 12.25 | 12.34 | 12.33 | 12.32 | 12.39 |
| 12.38 | 12.27 | 12.35 | 12.35 | 12.34 | 12.36 | 12.36 | 12.32 |
| 12.31 | 12.35 | 12.29 | 12.30 | | | | |

13) The results shown below were obtained by measuring the diameters of electric motor shafts produced on a centre lathe. Construct a frequency distribution and hence draw a histogram.

The measurements are given in 0.01 mm above 15.00 mm.

| | | | | | | | | | | | | | | | | | | |
|---|---|---|---|---|---|---|---|---|---|---|---|---|---|---|---|---|---|---|
| 5, | 8, | 3, | 5, | 5, | 6, | 3, | 0, | 6, | 2, | 5, | 2, | 2, | 1, | 9, | 6, | 8, | 6, | 3, | 6, |
| 4, | 2, | 7, | 4, | 5, | 4, | 5, | 8, | 2, | 8, | 8, | 5, | 2, | 5, | 7, | 5, | 3, | 7, | 6, | 2, |
| 8, | 6, | 6, | 3, | 5, | 5, | 7, | 9, | 6, | 4, | 8, | 3, | 2, | 6, | 4, | 9, | 2, | 8, | 6, | 4, |
| 6, | 3, | 1, | 8, | 4, | 2, | 4, | 1, | 1, | 9, | 5, | 7, | 7, | 5, | 5, | 7, | 2, | 6, | 2, | 4 |

14) Which of the following are discrete variables and which are continuous variables:
(a)  The size of men's shirts.
(b)  The length of plastic rod being produced in quantity.
(c)  The masses of castings being produced in a foundry.
(d)  The temperature of a furnace.
(e)  The number of electric motors produced per day in a factory.

# CUMULATIVE FREQUENCIES

| Crushing strength $(N/mm^2)$ | Frequency |
|---|---|
| 11.97 | 5 |
| 11.98 | 9 |
| 11.99 | 19 |
| 12.00 | 25 |
| 12.01 | 18 |
| 12.02 | 4 |

The measurements given in the table above may be arranged as a cumulative frequency distribution by adding each frequency to the total of the previous ones as shown in Table 3.

## TABLE 3

| Crushing strength $(N/mm^2)$ | Cumulative frequency |
|---|---|
| not more than 11.975 | 5 |
| not more than 11.985 | 5 + 9 = 14 |
| not more than 11.995 | 14 + 19 = 33 |
| not more than 12.005 | 33 + 25 = 58 |
| not more than 12.015 | 58 + 18 = 76 |
| not more than 12.025 | 76 + 8 = 80 |

A diagram showing the cumulative frequencies (Fig. 8.3) is called an ogive. Note that in arranging the cumulative frequency distribution we have used the upper boundary limit for each of the classes. We could have used the lower boundaries as shown in Table 4.

TABLE 4

| Crushing strength $(N/mm^2)$ | Cumulative frequency |
|---|---|
| less than 11.965 | 0 |
| less than 11.975 | 5 |
| less than 11.985 | $5 + 9 = 14$ |
| less than 11.995 | $14 + 19 = 33$ |
| less than 12.005 | $33 + 25 = 58$ |
| less than 12.015 | $58 + 18 = 76$ |
| less than 12.025 | $76 + 4 = 80$ |

Clearly the ogive will be the same as the one drawn in Fig. 8.3.

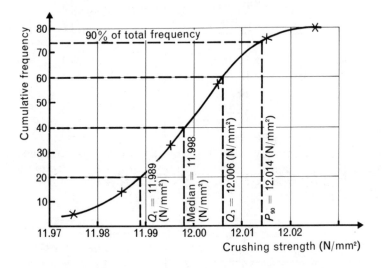

Fig. 8.3*

*For explanation of $Q_1$ and $Q_3$ in Fig. 8.3 see page 111 under 'Quartiles' and for the explanation of $P_{90}$ see page 111 under 'Percentiles'.

# THE MEDIAN

If a set of numbers is arranged in ascending (or descending) order of size, the median is the value which lies half-way along the set. Thus for the set

$$3, 4, 4, 5, 6, 7, 7, 9, 10$$

the median is 6

If there is an even number of values the median is found by taking the average of the two middle values. Thus for the set

$$3, 3, 5, 7, 9, 10, 13, 15$$

the median is $\frac{1}{2}(7 + 9) = 8$

**EXAMPLE 8.1**

The hourly wages of five employees in an office are £1.52, £2.96, £2.28, £8.20 and £2.75.

Arranging the amounts in ascending order we have

$$£1.52, \quad £2.28, \quad £2.75, \quad £2.96, \quad £8.20$$

The median is therefore £2.75.

# THE MEDIAN OF A CONTINUOUS FREQUENCY DISTRIBUTION

For a frequency distribution the median is the value of the variable corresponding to half the total frequency. Thus for the distribution of Table 3, the total frequency is 80 and the median is the value of the variable corresponding to a cumulative frequency of 40.

One way of finding the median is to use the ogive which represents the cumulative frequency distribution. Thus from Fig. 8.3, the median for the cumulative frequency distribution shown in Tables 3 and 4 is 11.998 mm.

EXAMPLE 8.2

Estimate the median of the frequency distribution shown below.

| Length (mm) | Frequency |
|:-----------:|:---------:|
| 798–800 | 7 |
| 801–803 | 38 |
| 804–806 | 88 |
| 807–809 | 54 |
| 810–812 | 13 |

Drawing up a cumulative frequency distribution we have:

| Length (mm) | Cumulative frequency |
|:-----------:|:--------------------:|
| not more than 800.5 | 7 |
| not more than 803.5 | $7 + 38 = \phantom{0}45$ |
| not more than 806.5 | $45 + 88 = 133$ |
| not more than 809.5 | $133 + 54 = 187$ |
| not more than 812.5 | $187 + 13 = 200$ |

Since the total frequency is 200 the median will be the value of the variable corresponding to a cumulative frequency of 100. Drawing the ogive (Fig. 8.4) we find the median to be 805 mm approximately.

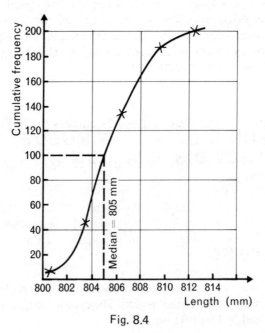

Fig. 8.4

# QUARTILES

We have seen that if a set of values is arranged in ascending or descending order the middle values (or the mean of the two middle values) which divides the set of values into two equal parts if the median. By extending this idea we can think of those values which divide the set into four equal parts. These values, usually denoted by $Q_1$, $Q_2$ and $Q_3$, are called the first, second and third quartiles. The value of $Q_2$ is equal to the median.

### EXAMPLE 8.3

Find the values of the first, second and third quartiles for the following: 3, 7, 9, 11, 4, 6, 7, and 10

Arranging the set in ascending order we have:

$$3, \ 4, \ 6, \ 7, \ 7, \ 9, \ 10, \ 11$$

The first quartile lies half way between the 2nd and 3rd values; the second quartile lies between the 4th and 5th terms; the third quartile lies between the 6th and 7th terms. Hence, $Q_1 = 5$, $Q_2 = 7$ and $Q_3 = 9.5$

For a frequency distribution the first quartile is the value of the variable corresponding to one-quarter of the total frequency. Likewise the third quartile is the value of the variable corresponding to three-quarters of the total frequency. The quartiles can therefore be estimated from an ogive which represents the cumulative frequency distribution.

### EXAMPLE 8.4

Find the values of the first and third quartiles for the distribution shown in Table 3.

Using the ogive drawn in Fig. 8.3, $Q_1 = 11.989 \, \text{N/mm}^2$, $Q_3 = 12.006 \, \text{N/mm}^2$.

# PERCENTILES

The percentiles are the values of the variable which divide the distribution into 100 equal parts. They are usually denoted by $P_1, P_2, P_3, \ldots, P_{99}$.

EXAMPLE 8.5

Find the value of the 90th percentile for the distribution shown in Table 3.

Using the ogive drawn in Fig. 8.3, the 90th percentile is the value of the variable corresponding to 90% of the total frequency, i.e. 90% of 80 = 72. Hence the value of the 90th ($P_{90}$) percentile is 12.014 N/mm² approximately.

# ARITHMETIC MEAN

The arithmetic mean is found by adding up all the observations in a set and dividing the result by the number of observations. That is,

$$\text{Arithmetic mean} = \frac{\text{Sum of the observations}}{\text{Number of observations}}$$

EXAMPLE 8.6

Five turned bars are measured and their diameters were found to be: 15.03, 15.02, 15.02, 15.00 and 15.03 mm. What is their mean diameter?

$$\text{Mean diameter} = \frac{15.03 + 15.02 + 15.02 + 15.00 + 15.03}{5}$$

$$= \frac{75.10}{5} = 15.02 \, \text{mm}$$

# THE MEAN OF A FREQUENCY DISTRIBUTION

The mean of a frequency distribution must take into account the frequencies as well as the measured observations.

If $x_1, x_2, x_3 \ldots x_n$ are measured observations which have frequencies $f_1, f_2, f_3 \ldots f_n$ then the mean of the distribution is

$$\bar{x} = \frac{x_1 f_1 + x_2 f_3 + x_3 f_3 + \ldots + x_n f_n}{f_1 + f_2 + f_3 \ldots + f_n} = \frac{\Sigma \, xf}{\Sigma \, f}$$

The symbol $\Sigma$ simply means the 'sum of'. Thus $\Sigma xf$ tells us to multiply together corresponding values of $x$ and $f$ and add the result together.

**EXAMPLE 8.7**

5 castings have a mass of 20.01 kg each, 3 have mass of 19.98 kg each and 2 have a mass of 20.03 kg each. What is the mean mass of the 10 castings?

The total mass is

$$(5 \times 20.01) + (3 \times 19.98) + (2 \times 20.03) = 200.05 \text{ kg}$$

$$\text{Mean mass} = \frac{\text{Total mass of the castings}}{\text{Number of castings}} = \frac{200.05}{10} = 20.005 \text{ kg}$$

# THE CODED METHOD FOR CALCULATING THE MEAN

The calculation of the mean may be speeded up considerably by using a unit method which is often referred to as using a coded method. The first step is to choose any value in the $x$ column to use as a datum for determining the coded values. A column may then be drawn up containing the actual values of $x$ in terms of units above or below the chosen value of $x$.

**EXAMPLE 8.8**

Find the mean of the frequency distribution shown below:

| $x$ | 14.96 | 14.97 | 14.98 | 14.99 | 15.00 | 15.01 | 15.02 | 15.03 | 15.04 |
|---|---|---|---|---|---|---|---|---|---|
| $f$ | 2 | 4 | 11 | 20 | 23 | 21 | 9 | 8 | 2 |

The best way of setting out the work is to make a table as shown opposite.

$$\text{Chosen value of } x = 15.00 \qquad \text{Unit size} = 0.01$$

Any value of $x$ may be chosen as the datum but the arithmetic will be simpler if a central value of $x$ is chosen.

The coded value for $x = 14.96$ is $-4$ because this value of $x$ is 4 units *less* than the chosen value of $x$. Similarly the coded value for $x = 15.02$ is $+2$ because 15.02 is 2 units *greater* than the chosen value of $x$. It is very important to assign to the coded value a plus or a minus sign depending on whether it is greater or less than the chosen value of $x$.

The mean may now be calculated from the coded values as follows:

| $x$ | $x_c$ | $f$ | $x_c f$ |
|---|---|---|---|
| 14.96 | $-4$ | 2 | $-8$ |
| 14.97 | $-3$ | 4 | $-12$ |
| 14.98 | $-2$ | 11 | $-22$ |
| 14.99 | $-1$ | 20 | $-20$ |
| 15.00 | 0 | 23 | 0 |
| 15.01 | 1 | 21 | 21 |
| 15.02 | 2 | 9 | 18 |
| 15.03 | 3 | 8 | 24 |
| 15.04 | 4 | 2 | 8 |
| | | 100 | $+9$ |
| | | $= \Sigma f$ | $= \Sigma x_c f$ |

$$\bar{x}_c = \frac{\Sigma x_c f}{\Sigma f} = \frac{9}{100} = 0.09$$

Actual value of $\bar{x}$ = (Chosen value of $x$) + ($\bar{x}_c \times$ Unit size)

$$= 15.00 + 0.09 \times 0.01 = 15.00 + 0.0009$$

$$= 15.0009$$

# MEAN OF A GROUPED DISTRIBUTION

The mean of a grouped distribution is found by taking the value of $x$ as the mid-points of the class intervals. Again, it is best to use the coded method as shown in Example 8.9.

## EXAMPLE 8.9

Find the mean of the grouped distribution shown in the table below.

| Diameter (mm) | 7.45-7.47 | 7.48-7.50 | 7.51-7.53 |
|---|---|---|---|
| Frequency | 16 | 34 | 28 |

| Diameter (mm) | 7.54-7.56 | 7.57-7.59 |
|---|---|---|
| Frequency | 18 | 4 |

Chosen value of $x = 7.52$ mm    Unit size $= 0.01$ mm

| Diameter (mm) | $x$ | $x_c$ | $f$ | $x_c f$ |
|---|---|---|---|---|
| 7.45–7.47 | 7.46 | −6 | 16 | −96 |
| 7.48–7.50 | 7.49 | −3 | 34 | −102 |
| 7.51–7.53 | 7.52 | 0 | 28 | 0 |
| 7.54–7.56 | 7.55 | +3 | 18 | +54 |
| 7.57–7.59 | 7.58 | +6 | 4 | +24 |
| | | | 100 | −120 |

$$x_c = \frac{\Sigma x_c f}{\Sigma f} = \frac{-120}{100} = -1.2$$

Since $\bar{x}_c$ is negative it indicates that the mean is 1.2 units less than the chosen size. That is,

$$\bar{x} = 7.52 - (1.2 \times 0.01) = 7.52 - 0.012 = 7.508 \text{ mm}$$

# THE MODE

The mode of a set of numbers is the number which occurs most frequently. Thus the mode of

$$2 \ 3 \ 3 \ 4 \ 4 \ 4 \ 5 \ 5 \ 6 \ 6 \ 7 \ 8$$

is 4, since this number occurs three times which is more than any of the other numbers in the set.

For a set of numbers the mode may not exist. Thus the set of numbers

$$4 \ 5 \ 6 \ 8 \ 9 \ 10 \ 12$$

has no mode.

It is possible for there to be more than one mode. The set of numbers

$$2 \ 3 \ 3 \ 5 \ 5 \ 5 \ 6 \ 6 \ 7 \ 8 \ 8 \ 8 \ 9 \ 10$$

has two modes, 5 and 8. The set of numbers is said to be *bimodal*. If there is only one mode, then the set of numbers is said to be *unimodal*.

# THE MODE OF A FREQUENCY DISTRIBUTION

The mode of a frequency distribution may be found by drawing a histogram as shown in Example 8.10.

### EXAMPLE 8.10

The table below shows the distribution of maximum loads supported by certain cables produced by the Steel Wire Company. Draw a histogram of this information and hence find the mode of the distribution.

| Maximum load (kN) | Number of cables |
|---|---|
| 84–88 | 4 |
| 89–93 | 10 |
| 94–98 | 24 |
| 99–103 | 34 |
| 104–108 | 28 |
| 109–113 | 12 |
| 114–118 | 6 |
| 119–123 | 2 |

Assuming that the measurements of load were accurate only to the nearest kilonewton, the class boundaries are

83.5–88.5, 88.5–93.5, 93.5–98.5, etc.

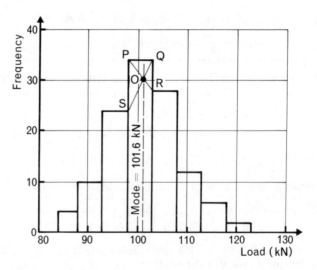

Fig. 8.5

The histogram is then as shown in Fig. 8.5. The mode is then found by drawing the diagonals PR and QS whose intersection is at O. The modal value is the value of the load corresponding to the point O. From the diagram this is found to be 101.6 kN.

# DISCUSSION ON THE MEAN, MEDIAN AND MODE

The arithmetic mean is the most familiar kind of average and it is extensively used in statistical work. However, in some cases the mean is definitely misleading. Again the mean size of screws is not of much use to the purchasing officer, because it might be at some point between stock sizes. In such cases the mode is probably the best value to use. However, which average is used will depend upon the particular circumstances.

## Exercise 8.2

1) Find the mode of the following set of numbers:

$$3, 5, 2, 7, 5, 8, 5, 2, 7, 6$$

2) Find the mode of: 38.7, 29.6, 32.1, 35.8, 43.2

3) Find the modes of: 8, 4, 9, 3, 5, 3, 8, 5, 3, 8, 9, 5, 6, 7

4) The data below relates to the resistance in ohms of an electrical part. Find the mode of this distribution, by drawing a histogram.

| Resistance (ohms) | 119 | 120 | 121 | 122 | 123 | 124 |
|---|---|---|---|---|---|---|
| Frequency | 5 | 9 | 19 | 25 | 18 | 4 |

5) Find the mode of the frequency distribution given in Question 18.

6) The information below shows the distribution of the diameters of rivet heads for rivets manufactured by a certain company.

| Diameter (mm) | 18.407–18.412 | 18.413–18.418 | 18.419–18.424 |
|---|---|---|---|
| Frequency | 2 | 6 | 8 |

| Diameter (mm) | 18.425–18.430 | 18.431–18.436 | 18.437–18.442 |
|---|---|---|---|
| Frequency | 12 | 7 | 3 |

Find the mode of this distribution:

(a) by drawing a histogram, (b) by calculation.

7) Find the median of the following set of numbers:

$$9, 2, 7, 3, 8, 5, 4$$

8) A student receives the following marks in an examination in five subjects: 84, 77, 95, 80 and 97. What is the median mark?

9) The following are the weekly wages earned by six people working in a small factory: £38, £71, £59, £63, £58 and £68. What is the median wage?

10) Find the mean and the median for the following set of observations: 15.63, 14.95, 16.00, 12.04, 15.88 and 16.04 ohms. Which of the two, the median or the mean, is, in your opinion, the better to use for these observations?

11) Draw up a cumulative frequency distribution for the distribution given in Table 5. Hence find the median and the 1st and 3rd quartiles for the distribution.

TABLE 5

| Resistance (ohms) | Frequency |
|---|---|
| 115 | 3 |
| 116 | 7 |
| 117 | 12 |
| 118 | 20 |
| 119 | 15 |
| 120 | 8 |
| 121 | 2 |

12) By drawing an ogive for the distribution of Table 6 determine the values of $Q_1$, $Q_2$ and $Q_3$.

TABLE 6

| Diameter (mm) | Frequency |
|---|---|
| 20.00–20.03 | 4 |
| 20.04–20.07 | 12 |
| 20.08–20.11 | 23 |
| 20.12–20.15 | 11 |
| 20.16–20.19 | 2 |

**13)** Find the lower and upper quartiles for the following set of values:

$$9, \ 8, \ 7, \ 4, \ 10, \ 11, \ 3, \ 4, \ 11, \ 13, \ 12, \ 12$$

**14)** The diameters of eight pipes were measured with the following results: 109.23, 109.21, 108.98, 109.03, 108.98. 109.22, 109.20, 108.91 mm. What is the mean diameter of the pipes?

**15)** 22 bricks have a mean mass of 24 kg and 18 similar bricks have a mean mass of 23.7 kg. What is the mean mass of the 40 bricks?

**16)** A sample of 100 lengths of timber was measured with the following results:

| Length (m) | 9.61 | 9.62 | 9.63 | 9.64 | 9.65 | 9.66 | 9.67 | 9.68 | 9.69 |
|---|---|---|---|---|---|---|---|---|---|
| Frequency | 2 | 4 | 12 | 18 | 31 | 22 | 8 | 2 | 1 |

Calculate the mean length of the timber.

**17)** The table below shows the distribution of the maximum loads supported by certain cables.

| Max. load (kN) | 19.2-19.5 | 19.6-19.9 | 20.0-20.3 | 20.4-20.7 |
|---|---|---|---|---|
| Frequency | 4 | 12 | 18 | 3 |

Calculate the mean load which the cables will support.

**18)** The table below shows a frequency distribution for the life-time of electric lamp bulbs.

| Lifetime (hours) | 400–499 | 500–599 | 600–699 | 700–799 | 800–899 |
|---|---|---|---|---|---|
| Frequency | 14 | 50 | 82 | 46 | 8 |

Calculate the mean lifetime of the electric light bulbs.

# FREQUENCY CURVES

A frequency curve may be drawn by joining the mid-points of the top of the rectangles in a histogram. The frequency curve is a convenient way of representing frequency distributions to make comparisons between them easier (Fig. 8.6).

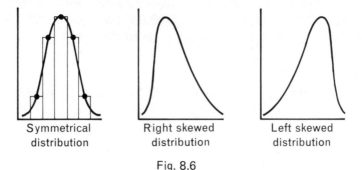

| Symmetrical distribution | Right skewed distribution | Left skewed distribution |

Fig. 8.6

# MEASURES OF DISPERSION

The central tendency of a distribution, as given by the mean, mode or median, gives some idea about the position of the distribution from the reference axis. For instance, Fig. 8.7 shows three similar distributions which have different means.

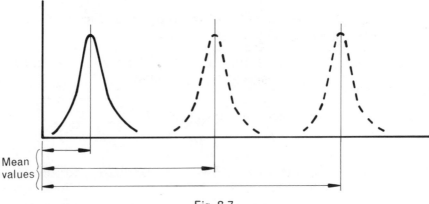

Mean values

Fig. 8.7

However, Fig. 8.8 shows two different distributions which have the same mean but very different spread or dispersion. We need, therefore a measure which will define this spread or dispersion. The measures used are the *range* and the *standard deviation.*

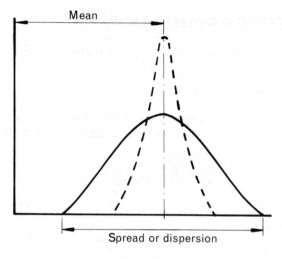

Fig. 8.8

# THE RANGE

The range is the difference between the largest observation in a set and the smallest observation in the set. That is,

Range = Largest observation − Smallest observation

The range gives some idea of the spread of the distribution but it depends solely upon the end values. It gives no indication of the distribution of the data and hence it is never used as a measure of dispersion for a frequency distribution. However, if only a small number of observations are taken as was the case in Example 8.11, the range is a very effective measure of dispersion.

### EXAMPLE 8.11

The power of five similar electric fires were measured with the following results: 1010, 1030, 998, 999, 1000 W. Find the range for these five measurements.

Smallest measurement = 998 W

Largest measurement = 1030 W

∴ Range = 1030 − 998 = 32 W

# THE STANDARD DEVIATION

The most valuable and widely used measure of dispersion is the standard deviation. It is always represented by the Greek letter $\sigma$ (sigma) and it may be calculated from the formula,

$$\sigma = \sqrt{\frac{\Sigma x^2 f}{\Sigma f} - \bar{x}^2}$$

The best way of calculating the standard deviation is to use a coded method as shown in Example 8.12.

### EXAMPLE 8.12

Calculate the mean and standard deviation for the following frequency distribution.

| Resistance (ohm) | 5.37 | 5.38 | 5.39 | 5.40 | 5.41 | 5.42 | 5.43 | 5.44 |
|---|---|---|---|---|---|---|---|---|
| Frequency | 4 | 10 | 14 | 24 | 34 | 18 | 10 | 6 |

Chosen value of $x = 5.40$     Unit size $= 0.01$ ohm

| $x$ | $x_c$ | $f$ | $x_c f$ | $x_c^2 f$ |
|---|---|---|---|---|
| 5.37 | $-3$ | 4 | $-12$ | 36 |
| 5.38 | $-2$ | 10 | $-20$ | 40 |
| 5.39 | $-1$ | 14 | $-14$ | 14 |
| 5.40 | 0 | 24 | 0 | 0 |
| 5.41 | 1 | 34 | 34 | 34 |
| 5.42 | 2 | 18 | 36 | 72 |
| 5.43 | 3 | 10 | 30 | 90 |
| 5.44 | 4 | 6 | 24 | 96 |
| | | 120 | 78 | 382 |

Now $\bar{x}_c = \dfrac{\Sigma x_c f}{\Sigma f} = \dfrac{78}{120} = 0.65$

$\therefore \quad \bar{x} = 5.40 + 0.65 \times 0.01 = 5.4065$ ohm

and $\sigma_c = \sqrt{\dfrac{\Sigma x_c^2 f}{\Sigma f} - (\bar{x}_c)^2} = \sqrt{\dfrac{382}{120} - (0.65)^2} = \sqrt{2.7068}$

$\qquad = 1.662$

$\therefore \quad \sigma = \sigma_c \times \text{unit size} = 1.662 \times 0.01 = 0.016\,62$ ohm

# Rough Check on Standard Deviation using Range

If the distribution is reasonably symmetrical, as is the one in Example 8.12, a rough check for the standard deviation may be obtained by finding the range of the data and dividing it by 6. Thus for Example 8.12,

$$\text{Range} = 5.44 - 5.37 = 0.07$$

$$\sigma \text{ (roughly)} = \frac{0.07}{6} = 0.012 \text{ ohm}$$

The calculated value of 0.016 57 is therefore of the right order (i.e. it is not wildly incorrect).

### EXAMPLE 8.13

The table indicates experimental results from a sample of resistors.

| Resistance (ohm) | 24.92–24.94 | 24.95–24.97 | 24.98–25.00 |
|---|---|---|---|
| Frequency | 2 | 3 | 9 |

| Resistance (ohm)· | 25.01–25.03 | 25.04–25.06 | 25.07–25.09 |
|---|---|---|---|
| Frequency | 23 | 18 | 5 |

Calculate the standard deviation.

Chosen value of $x = 25.02$ ohm       Unit size $= 0.03$ ohm

| Class | $x$ | $x_c$ | $f$ | $x_c f$ | $x_c^2 f$ |
|---|---|---|---|---|---|
| 24.92–24.94 | 24.93 | −3 | 2 | −6 | 18 |
| 24.95–24.97 | 24.96 | −2 | 3 | −6 | 12 |
| 24.98–25.00 | 24.99 | −1 | 9 | −9 | 9 |
| 25.01–25.03 | 25.02 | 0 | 23 | 0 | 0 |
| 25.04–25.06 | 25.05 | 1 | 18 | 18 | 18 |
| 25.07–25.09 | 25.08 | 2 | 5 | 10 | 20 |
| | | | 60 | 7 | 77 |

A unit size of 0.03 ohm has been chosen because each value of $x$ differs from its preceding value by 0.03. Making the unit size as large as possible simplifies the calculation of the standard deviation.

Now    $\bar{x}_c = \dfrac{+7}{60} = +0.116$

∴    $\bar{x} = 25.02 + (0.03)(0.116) = 25.0235\,\text{ohm}$

Also    $\sigma_c = \sqrt{\dfrac{77}{60} - (0.116)^2} = \sqrt{1.2697} = 1.1268$

∴    $\sigma = 1.1268 \times 0.03 = 0.0338\,\text{ohm}$

(A rough check for $\sigma$ gives $\dfrac{25.09 - 24.92}{6} = 0.0283$, which is of the same order as the value calculated above.)

# Exercise 8.3

1) Calculate the range, mean and standard deviation for the following five observations: 16.01, 16.00, 15.98, 15.97 and 15.99 mm.

2) Calculate the mean and standard deviation for the following frequency distribution which relates to the compressive strength of load-carrying bricks.

| Strength (N/mm²) | 11.46 | 11.47 | 11.48 | 11.49 | 11.50 | 11.51 | 11.52 | 11.53 |
|---|---|---|---|---|---|---|---|---|
| Frequency | 1 | 4 | 12 | 15 | 11 | 6 | 3 | 1 |

3) In water absorption tests on 100 bricks the following results were obtained:

| % Absorption | 7 | 8 | 9 | 10 | 11 | 12 | 13 | 14 |
|---|---|---|---|---|---|---|---|---|
| Frequency | 1 | 4 | 9 | 24 | 30 | 26 | 5 | 1 |

Calculate the mean and standard deviation.

4) 100 watts is the nominal value of the sample of electric light bulbs tabulated below. Calculate the mean and standard deviation of the actual correct values.

| Power (watt) | 99.6 | 99.7 | 99.8 | 99.9 | 100.0 | 100.1 | 100.2 | 100.3 |
|---|---|---|---|---|---|---|---|---|
| Frequency | 3 | 8 | 13 | 18 | 15 | 9 | 6 | 3 |

**5)** A brand of washing powder tested prior to marketing revealed its capacity to launder woollen garments of equivalent sizes.

| Laundered garments | 5-7 | 8-10 | 11-13 | 14-16 | 17-19 |
|---|---|---|---|---|---|
| Frequency | 1 | 5 | 11 | 7 | 3 |

From the data above calculate the mean and standard deviation.

**6)** It is considered that a person on average uses 40 gallons of water daily. Using the following tabulated data calculate the mean water consumption and the standard deviation.

| Daily consumption (gallons) | 30-34 | 35-39 | 40-44 | 45-49 | 50-54 | 55-59 |
|---|---|---|---|---|---|---|
| Number of users | 3 | 19 | 43 | 26 | 7 | 2 |

# AREAS AND VOLUMES

1. *Calculate the volumes and surface areas of pyramids, cones and spheres.*
2. *Define frustum.*
3. *Calculate the surface area and volume of frusta of cones and pyramids.*
4. *Identify the component basic shapes composite areas and volumes (i.e. rectangles, triangles, circles, prisms, pyramids, cones, cylinders and spheres).*
5. *Calculate the total area or volume of composite figures.*
6. *Use the formula $\pi ab$ and $\pi(a+b)$ for the area and approximate perimeter of an ellipse.*
7. *Use the prismoidal rule to calculate volume where appropriate.*
8. *Calculate areas of irregular shapes using the mid-ordinate, trapezoidal and Simpson's rule.*
9. *State numerical results of any calculation in this chapter to an accuracy consistent with the data used.*

## UNITS OF AREA

The standard abbreviation for units of area are:

$$\text{Square metres } = \text{ m}^2$$

$$\text{Square millimetres } = \text{ mm}^2$$

Conversion of square units of area are:

$$1\,\text{m}^2 \ = \ (1000\,\text{mm})^2 \ = \ (1000 \times 1000)\,\text{mm}^2 \ = \ 10^6\,\text{mm}^2$$

For large areas the hectare is used such that:

$$1\,\text{hectare (ha) } = \ 10000\,\text{m}^2$$

## AREAS AND PERIMETERS

### Rectangle

$$\text{Area } = \ l \times b$$
$$\text{Perimeter } = \ 2l + 2b$$

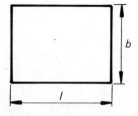

**EXAMPLE 9.1**

Find the area of the section shown in Fig. 9.1.

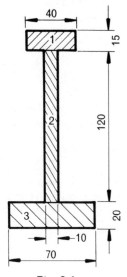

The section can be split up into three rectangles as shown. The total area can be found by calculating the areas of the three rectangles separately and then adding these together. Thus,

Area of rectangle 1  $= 15 \times 40$
$= 600 \, \text{mm}^2$

Area of rectangle 2  $= 10 \times 120$
$= 1200 \, \text{mm}^2$

Area of rectangle 3  $= 20 \times 70$
$= 1400 \, \text{mm}^2$

Total area of section  $= 600 + 1200$
$+ 1400 = 3200 \, \text{mm}^2$

Fig. 9.1

# Parallelogram

$$\boxed{\text{Area} = b \times h}$$

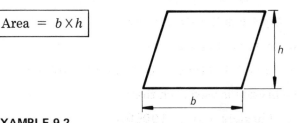

**EXAMPLE 9.2**

Find the area of the parallelogram shown in Fig. 9.2.

The first step is to find the vertical height $h$.

In $\triangle \text{BCE}$,

$h = \text{BC} \times \sin 60° = 3 \times 0.866 = 2.598$

$\left( \begin{array}{c} \text{Area of} \\ \text{parallelogram} \end{array} \right) = \begin{array}{l} \text{Base} \times \\ \text{Vertical height} \end{array}$

$= 5 \times 2.598$

$= 13.0 \, \text{m}^2$

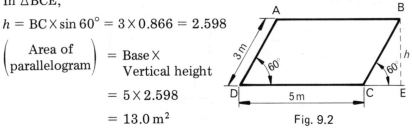

Fig. 9.2

# Triangle

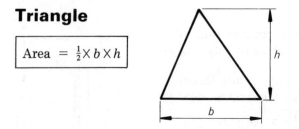

$$\boxed{\text{Area} = \tfrac{1}{2} \times b \times h}$$

## EXAMPLE 9.3

The inner shape of a pattern in a large hall is a regular octagon (8-sided polygon) which is 5 m across flats (Fig. 9.3). Find its area.

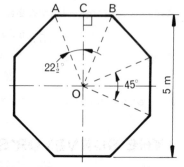

The angle subtended at the centre by a side of the octagon $= \dfrac{360°}{8} = 45°$.

Now triangle AOB is isosceles, since OA = OB.

Fig. 9.3

∴       $\angle AOC = \dfrac{45°}{2} = 22°30'$

But       $OC = \dfrac{5}{2} = 2.5\,\text{m}$

Also       $\dfrac{AC}{OC} = \tan 22°30'$

∴       $AC = OC \times \tan 22°30' = 2.5 \times 0.4142 = 1.035\,\text{m}$

Thus    Area of $\triangle AOB = AC \times OC = 1.035 \times 2.5 = 2.588\,\text{m}^2$

∴      Area of octagon $= 2.588 \times 8 = 20.7\,\text{m}^2$

# Trapezium (or Trapezoid)

$$\boxed{\text{Area} = \tfrac{1}{2} \times h \times (a + b)}$$

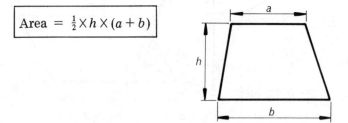

**EXAMPLE 9.4**

Fig. 9.4 shows the cross-section of a retaining wall. Calculate its cross-sectional area.

Since the section is a trapezium:

$$\text{Area} = \tfrac{1}{2} \times h \times (a + b)$$

$$= \tfrac{1}{2} \times 6 \times (2 + 3)$$

$$= \tfrac{1}{2} \times 6 \times 5$$

$$= 15 \, \text{m}^3$$

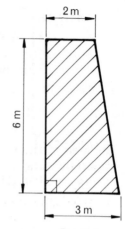

Fig. 9.4

# THE SURVEYOR'S FIELD BOOK

Fig. 9.5 shows the measurements made when surveying a plot of land. The offsets CK, GF, BH and JE are measured perpendicularly to the diagonal AD and the distances AG, AH, AK and AD are also measured. Hence the vertices A, B, C, D, E and F are fixed.

In recording the information in the surveyor's field book, distances from one end of the base line AD are written down in order in the centre column and the offsets are recorded to the right or left of the centre column as shown opposite.

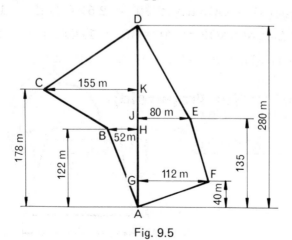

Fig. 9.5

|          | 280 (D)<br>178 (K) |          |
|----------|--------------------|----------|
| 155 (C)  | 135 (J)            | 80 (E)   |
| 52 (B)   | 122 (H)            |          |
|          | 40 (G)             | 112 (F)  |
|          | From A<br>in metres |          |

The area of the plot is now found by considering the areas of the trapeziums and triangles formed by the base line AD and the offsets, CK, BH, GF and JE. Thus,

$$\text{Area of triangle AGF} \;=\; \tfrac{1}{2} \times 40 \times 112 \qquad\quad =\quad 2240\,\text{m}^2$$

$$\text{Area of trapezium GFEJ} \;=\; \tfrac{1}{2} \times 95 \times (80 + 112) \;=\; 9120\,\text{m}^2$$

$$\text{Area of triangle JED} \;=\; \tfrac{1}{2} \times 145 \times 80 \qquad\quad =\quad 5800\,\text{m}^2$$

$$\text{Area of triangle DCK} \;=\; \tfrac{1}{2} \times 102 \times 155 \qquad\; =\quad 7905\,\text{m}^2$$

$$\text{Area of trapezium CKHB} \;=\; \tfrac{1}{2} \times 56 \times (155 + 52) \;=\; 5796\,\text{m}^2$$

$$\text{Area of triangle ABH} \;=\; \tfrac{1}{2} \times 122 \times 52 \qquad\quad =\quad \underline{3172\,\text{m}^2}$$

$$\text{Hence the total area of the surveyed plot} \;=\; 34033\,\text{m}^2$$

## Circle

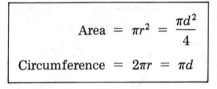
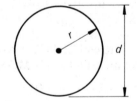

$$\text{Area} \;=\; \pi r^2 \;=\; \frac{\pi d^2}{4}$$

$$\text{Circumference} \;=\; 2\pi r \;=\; \pi d$$

**EXAMPLE 9.5**

A pipe has an outside diameter of 32.5 mm and an inside diameter of 25 mm. Calculate the cross-sectional area of the shaft (Fig. 9.6).

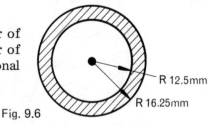

R 12.5mm

R 16.25mm

Fig. 9.6

$$\text{Area of cross-section} \;=\; \text{Area of outside circle} - \text{Area of inside circle}$$

$$=\; \pi \times 16.25^2 - \pi \times 12.5^2 \;=\; 338\,\text{mm}^2$$

# Sector of a Circle

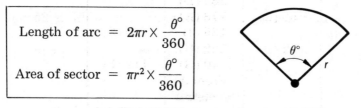

| |
|---|
| Length of arc $= 2\pi r \times \dfrac{\theta°}{360}$ |
| Area of sector $= \pi r^2 \times \dfrac{\theta°}{360}$ |

### EXAMPLE 9.6

Calculate **a)** the length of arc of a circle whose radius is 8 m and which subtends an angle of 56° at the centre, and **b)** the area of the sector so formed.

**a)** Length of arc $= 2\pi r \times \dfrac{\theta°}{360} = 2 \times \pi \times 8 \times \dfrac{56}{360} = 7.82\,\text{m}$

**b)** Area of sector $= \pi r^2 \times \dfrac{\theta°}{360} = \pi \times 8^2 \times \dfrac{56}{360} = 31.3\,\text{m}^2$

### EXAMPLE 9.7

In a circle of radius 40 mm a chord is drawn which subtends an angle of 120° at the centre. What is the area of the minor segment?

Area of sector MCNO

$= \dfrac{\pi r^2 \theta°}{360} = \dfrac{\pi \times 40^2 \times 120}{360} = 1676\,\text{mm}^2$

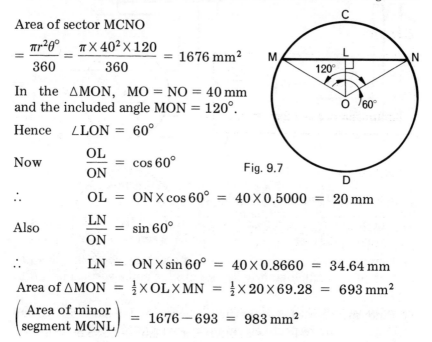

Fig. 9.7

In the $\triangle$MON, MO = NO = 40 mm and the included angle MON = 120°.

Hence  $\angle$LON $= 60°$

Now  $\dfrac{\text{OL}}{\text{ON}} = \cos 60°$

$\therefore$  OL $=$ ON $\times \cos 60° = 40 \times 0.5000 = 20\,\text{mm}$

Also  $\dfrac{\text{LN}}{\text{ON}} = \sin 60°$

$\therefore$  LN $=$ ON $\times \sin 60° = 40 \times 0.8660 = 34.64\,\text{mm}$

Area of $\triangle$MON $= \frac{1}{2} \times$ OL $\times$ MN $= \frac{1}{2} \times 20 \times 69.28 = 693\,\text{mm}^2$

$\left(\begin{array}{c}\text{Area of minor}\\\text{segment MCNL}\end{array}\right) = 1676 - 693 = 983\,\text{mm}^2$

## Exercise 9.1

1) The area of a metal plate is $220\,mm^2$. If its width is $25\,mm$, find its length.

2) A sheet metal plate has a length of $147.5\,mm$ and a width of $86.5\,mm$. Find its area in $m^2$.

3) Find the areas of the sections shown in Fig. 9.8.

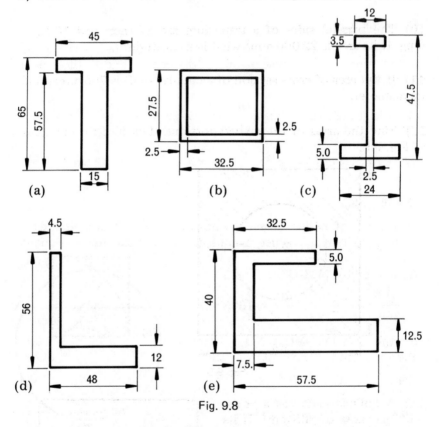

Fig. 9.8

4) What is the area of a parallelogram whose base is $70\,mm$ long and whose vertical height is $40\,mm$?

5) Obtain the area of a parallelogram if two adjacent sides measure $112.5\,mm$ and $105\,mm$ and the angle between them is $49°$.

6) Determine the length of the side of a square whose area is equal to that of a parallelogram with a $3\,m$ base and a vertical height of $1.5\,m$.

7) Find the area of a trapezium whose parallel sides are $75\,mm$ and $82\,mm$ long respectively and whose vertical height is $39\,mm$.

8) Find the area of a regular hexagon,

(a) which is 40 mm wide across flats,

(b) which has sides 50 mm long.

9) Find the area of a regular octagon,

(a) which is 2 mm wide across flats,

(b) which has sides 2 mm long.

10) The parallel sides of a trapezium are 120 mm and 160 mm long. If its area is 22 000 mm² what is its altitude?

11) If the area of cross-section of a circular shaft is 700 mm², find its diameter.

12) Find the areas of the shaded portions of each of the diagrams of Fig. 9.9.

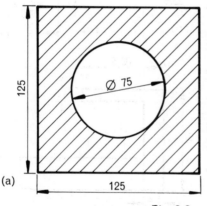
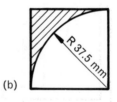

(a)

(b)

Fig. 9.9

13) A hollow shaft has a cross-sectional area of 868 mm². If its inside diameter is 7.5 mm, calculate its outside diameter.

14) Find the area of the blank shown in Fig. 9.10.

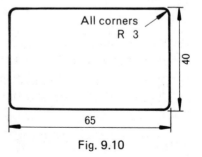

Fig. 9.10

15) How many revolutions will a wheel make in travelling 2 km if its diameter is 700 mm?

**16)** A bay window, semi-circular in plan, is to be covered with lead. Its radius is 2.4 m. Calculate:

(a) the area of lead required,

(b) the length of skirting board required to go round the bay.

**17)** A rectangular piece of insulating material is required to wrap round a pipe which is 560 mm diameter. Allowing 150 mm for overlap, calculate the width of material required.

**18)** The following figures are taken from a surveyor's field book. Roughly sketch the survey and find the area of the plot.

(a)

|  | To E | |
|---|---|---|
|  | 150 | |
|  | 90 | 26.4 to D |
| To F 32.0 | 78 | |
|  | 76 | 13.6 to C |
|  | 16 | 52.0 to B |
|  | From A | |

(b)

|  | To Y | |
|---|---|---|
|  | 50 | |
|  | 40 | 15 to C |
|  | 32 | 18 to D |
| To B 6 | 31 | |
| To A 6 | 24 | 25 to E |
|  | 19 | 15 to F |
|  | From X | |

**19)** The centre panel of a large ceiling consists of a 2.2 m diameter circle circumscribed by a regular hexagon. Find:

(a) the area of the circle,
(b) the area of the hexagon,
(c) the area between the circle and the hexagon.

**20)** The floor of a summer house is an octagon which could be inscribed inside a circle whose diameter is 7 m. Find the area of the floor.

# UNITS OF VOLUME OR CAPACITY

The capacity of a container is the volume that it will contain. It is often measured in the same units as volume, that is cubic metres.

Sometimes however, as in the case of liquid measure, the litre*
(abbreviation ℓ) unit is used such that

$$1 \, \text{m}^3 \ = \ 1000 \, \ell \quad \text{to four figure accuracy}$$

Small capacities are often measured in millilitres (mℓ) and

$$1000 \, \text{millilitres (m}\ell) \ = \ 1 \, \text{litre} \, (\ell) \quad \text{to four figure accuracy}$$

but there are $1\,000\,000 \, \text{mm}^3$ in 1 litre

$$\therefore \qquad\qquad 1 \, \text{m}\ell \ = \ 1000 \, \text{mm}^3 \quad \text{to four figure accuracy}$$

# VOLUMES AND SURFACE AREAS

## Any Solid Having a Uniform Cross-section and Parallel End Faces

Volume = Cross-sectional area × Length of solid

Surface area = Longitudinal surface + Ends
         i.e. (Perimeter of cross-section × Length of solid)
            + (Total area of ends)

A *prism* is the name often given this type of solid if the cross-
section is triangular or polygonal.

**EXAMPLE 9.8**

A piece of timber has the cross-section shown in Fig. 9.11. If its
length is 300 mm, find its volume and total surface area.

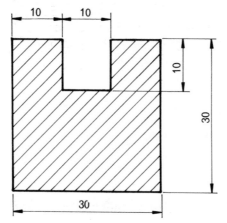

Fig. 9.11

*A litre is 1.000 028 cubic decimetres, but the term 'litre' is *not* to be used
for precise measurements.

Area of cross-section $= (30 \times 30) - (10 \times 10) = 800\,\mathrm{mm}^2$

Volume $=$ (Area of cross-section) $\times$ (Length)

$\qquad = 800 \times 300 = 240\,000\,\mathrm{mm}^3$

Perimeter of cross-section $= (3 \times 30) + (5 \times 10) = 140\,\mathrm{mm}$

Total surface area $=$ (Perimeter of cross-section $\times$ Length)
$\qquad\qquad\qquad + $ (Area of ends)

$\qquad\qquad = (140 \times 300) + (2 \times 800) = 43\,600\,\mathrm{mm}^2$

**EXAMPLE 9.9**

A steel section has the cross-section shown in Fig. 9.12. If it is 9 m long, calculate its volume and total surface area.

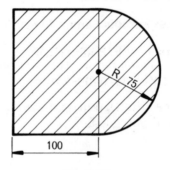

Fig. 9.12

To find the volume

$\qquad$ Area of cross-section $= \frac{1}{2} \times \pi \times 75^2 + 100 \times 150 = 23\,840\,\mathrm{mm}^2$

$\qquad\qquad\qquad\qquad = \dfrac{23\,840}{(1000)^2} = 0.023\,84\,\mathrm{m}^2$

$\therefore \qquad$ Volume of solid $= 0.023\,84 \times 9 = 0.215\,\mathrm{m}^3$

To find the surface area

$\qquad$ Perimeter of cross-section $= \pi \times 75 + 2 \times 100 + 150$

$\qquad\qquad\qquad\qquad = 585.5\,\mathrm{mm} = \dfrac{585.5}{1000} = 0.5855\,\mathrm{m}$

$\qquad$ Lateral surface area $= 0.5855 \times 9 = 5.270\,\mathrm{m}^2$

$\qquad$ Surface area of ends $= 2 \times 0.024 = 0.048\,\mathrm{m}^2$

$\therefore \qquad$ Total surface area $= 5.27 + 0.05 = 5.32\,\mathrm{m}^2$

# Cylinder

| |
|---|
| Volume $= \pi r^2 h$ |
| Surface area $= 2\pi rh + 2\pi r^2 = 2\pi r(h+r)$ |

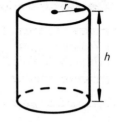

## EXAMPLE 9.10

A cylindrical can holds 18 litres of petrol. Find the depth of the petrol if the can has a diameter of 600 mm.

Now    18 litres $= 18 \times 10^6 \, \text{mm}^3$

and if the depth of the petrol is $h$ mm

then    Volume of petrol $= \pi(\text{Radius})^2 \times h$

$\therefore$ $\qquad\qquad 18 \times 10^6 = \pi \times 300^2 \times h$

$\therefore$ $\qquad\qquad h = \dfrac{18\,000\,000}{\pi \times 90\,000} = 63.7 \, \text{mm}$

## EXAMPLE 9.11

A metal bar of length 200 mm and diameter 75 mm is melted down and cast into washers 2.5 mm thick with an internal diameter of 12.5 mm and external diameter 25 mm. Calculate the number of washers obtained assuming no loss of metal.

Volume of original bar of metal $= \pi \times 37.5^2 \times 200$

$\qquad\qquad\qquad\qquad\qquad = 883\,500 \, \text{mm}^3$

Volume of one washer $= \pi \times (12.5^2 - 6.25^2) \times 2.5$

$\qquad\qquad\qquad\qquad = \pi \times 117.2 \times 2.5$

$\qquad\qquad\qquad\qquad = 920.4 \, \text{mm}^3$

Number of washers obtained $= \dfrac{883\,500}{920.4} = 960$

# Cone

| |
|---|
| Volume $= \frac{1}{3}\pi r^2 h$<br>($h$ is the vertical height) |
| Curved surface area $= \pi r l$<br>($l$ is the slant length) |

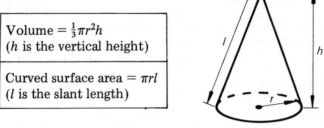

## EXAMPLE 9.12

A hopper is in the form of an inverted cone. It has a maximum internal diameter of 2.4 m and an internal height of 2.1 m.

**a)** If the hopper is to be lined with lead, calculate the area of lead required.

**b)** Determine the capacity of the hopper before lining.

**a)** The slant height may be found by using Pythagoras' theorem on the triangle shown in Fig. 9.13.

$$l^2 = r^2 + h^2 = 1.2^2 + 2.1^2 = 5.85$$
$$l = \sqrt{5.85} = 2.42\,\text{m}$$

Surface area $= \pi r l = \pi \times 1.2 \times 2.42 = 9.12\,\text{m}^2$

Hence the area of lead required is $9.12\,\text{m}^2$

**b)** Volume $= \frac{1}{3}\pi r^2 h = \frac{1}{3} \times \pi \times 1.2^2 \times 2.1 = 3.17\,\text{m}^3$

Hence the capacity of the hopper is $3.17\,\text{m}^3$

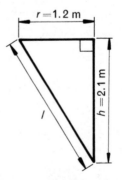

Fig. 9.13

# Frustum of a Cone

A *frustum* is the portion of a cone or pyramid between the base and a horizontal slice which removes the pointed portion.

Volume $= \frac{1}{3}\pi h(R^2 + Rr + r^2)$
($h$ is the vertical height)

Curved surface area $= \pi l(R + r)$
Total surface area $= \pi l(R + r) + \pi R^2 + \pi r^2$
($l$ is the slant height)

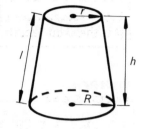

### EXAMPLE 9.13

A concrete column is shaped like a frustum of a cone. It is 2.8 m high. The radius at the top is 0.6 m and the base radius is 0.9 m. Calculate the volume of concrete in the column.

$$\begin{aligned}
\text{Volume} &= \tfrac{1}{3}\pi h(R^2 + Rr + r^2) \\
&= \tfrac{1}{3}\times \pi \times 2.8 \times (0.9^2 + 0.9 \times 0.6 + 0.6^2) \\
&= \tfrac{1}{3}\times \pi \times 2.8 \times 1.71 \ = \ 5.01\,\text{m}^3
\end{aligned}$$

### EXAMPLE 9.14

The bowl shown in Fig. 9.14 is made from sheet steel and has an open top. Calculate the total cost of painting the vessel (inside and outside) at a cost of 1 p per 10 000 mm².

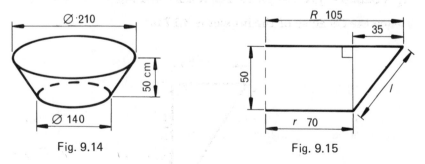

Fig. 9.14                    Fig. 9.15

Fig. 9.15 shows a half section of the bowl. Using Pythagoras' theorem on the right-angled triangle

$$l^2 = 50^2 + 35^2$$

$$\therefore \qquad l = 61.0\,\text{mm}$$

Now the required total surface area, i.e. inside and outside

$$= 2\{\text{Curved surface area}\} + 2(\text{Base area})$$
$$= 2\{\pi l(R + r)\} + 2(\pi r^2)$$
$$= 2\{\pi(61)(105 + 70)\} + 2(\pi 70^2) = 97\,800 \text{ mm}^2$$

At 1p per $10\,000 \text{ mm}^2$ total cost $= \dfrac{97\,800}{10\,000} = 9.78\text{p}$

## Sphere

| |
|---|
| Volume $= \frac{4}{3}\pi r^3$ |
| Surface area $= 4\pi r^2$ |

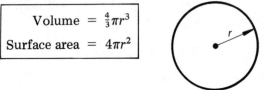

**EXAMPLE 9.15**

A concrete dome is in the shape of a hemisphere. Its internal and external diameters are 4.8 m and 5 m respectively.

a) Calculate the volume of concrete used in its construction.
b) If the inside is to be painted, calculate the area to be covered.

a) If $R$ is the outside radius, $r$ the inside radius and $V$ the volume, then

$$V = \tfrac{1}{2}(\tfrac{4}{3}\pi R^3 - \tfrac{4}{3}\pi r^3)$$
$$= \tfrac{1}{2} \times \tfrac{4}{3} \times \pi \times (R^3 - r^3)$$
$$= \tfrac{2}{3} \times \pi \times (5^3 - 4.8^3) = 30.2 \text{ m}^3$$

b)    Area of inside of dome $= \tfrac{1}{2} \times 4\pi r^2$
$$= \tfrac{1}{2} \times 4 \times \pi \times 4.8^2 = 145 \text{ m}^2$$

## Spherical Cap

$$\begin{pmatrix} \text{Total} \\ \text{surface area} \end{pmatrix} = \begin{pmatrix} \text{Curved} \\ \text{surface area} \end{pmatrix} + \begin{pmatrix} \text{Flat} \\ \text{base area} \end{pmatrix}$$
$$= 2\pi Rh + \pi r^2$$
$$\text{or } \pi(r^2 + h^2) + \pi r^2$$

$$\text{Volume} = \frac{\pi h^2}{3}(3R - h) \quad \text{or} \quad \frac{\pi h}{6}(3r^2 + h^2)$$

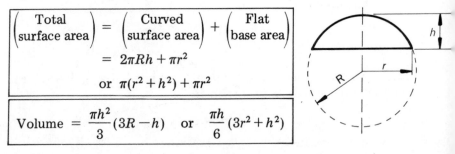

**EXAMPLE 9.16**

Calculate the volume and outer curved surface area of a spherical dome having a span of 12 m and a rise of 4 m.

Using the formula

$$V = \frac{\pi h}{6}(3r^2 + h^2)$$

when $r = 6$ and $h = 4$,   $V = \dfrac{\pi \times 4}{6} \times (3 \times 6^2 + 4^2)$

$$= 2.09 \times (108 + 16) = 260 \text{ m}^3$$

Using the formula      $A = \pi(r^2 + h^2)$

when $r = 6$ and $h = 4$,    $A = \pi(6^2 + 4^2) = 163 \text{ m}^2$

## Pyramid

Volume $= \frac{1}{3}Ah$

Surface area = Sum of the areas of the triangles forming the sides plus the area of the base

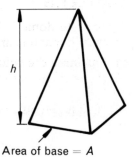

Area of base $= A$

**EXAMPLE 9.17**

Find the volume and total surface area of a symmetrical pyramid whose base is a rectangle $7 \text{ m} \times 4 \text{ m}$ and whose height is 10 m.

Base area $= 7 \times 4 = 28 \text{ m}^2$

Height   $= 10 \text{ m}$

Volume   $= \frac{1}{3}Ah = \frac{1}{3} \times 28 \times 10$

$$= 93.3 \text{ m}^3$$

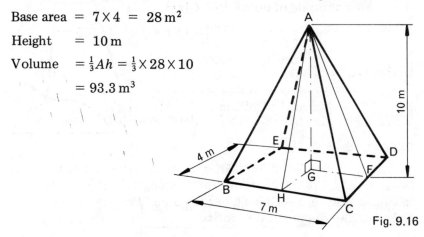

Fig. 9.16

From Fig. 9.16 the surface area consists of two sets of equal triangles (that is $\triangle ABC$ and $\triangle ADE$, and also $\triangle ABE$ and $\triangle ACD$) together with the base BCDE. To find the area of $\triangle ABC$ we must find the slant height AH. From the apex, A, drop a perpendicular AG on to the base and draw GH perpendicular to BC. H is then the mid-point of BC.

In $\triangle AHG$, $\angle AGH = 90°$ and, by Pythagoras' theorem,

$$AH^2 = AG^2 + HG^2 = 10^2 + 2^2 = 104$$

$\therefore \qquad AH = \sqrt{104} = 10.2\,\text{m}$

$\therefore \qquad$ Area of $\triangle ABC = \tfrac{1}{2} \times \text{Base} \times \text{Height}$

$$= \tfrac{1}{2} \times 7 \times 10.20 = 35.7\,\text{m}^2$$

Similarly, to find the area of $\triangle ACD$ we must find the slant height AF. Draw GF, F being the mid-point of CD. Then in $\triangle AGF$, $\angle AGF = 90°$ and by Pythagoras' theorem,

$$AF^2 = AG^2 + GF^2 = 10^2 + 3.5^2 = 112.3$$

$\therefore \qquad AF = \sqrt{112.3} = 10.6\,\text{m}$

$\therefore \qquad$ Area of $\triangle ACD = \tfrac{1}{2} \times \text{Base} \times \text{Height}$

$$= \tfrac{1}{2} \times 4 \times 10.6 = 21.2\,\text{m}^2$$

$\therefore \quad$ Total surface area $= (2 \times 35.7) + (2 \times 21.2) + (7 \times 4)$

$$= 14.2\,\text{m}^2$$

## Frustum of a Pyramid

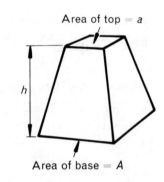

Area of top $= a$

Area of base $= A$

> Volume $= \tfrac{1}{3}h(A + \sqrt{Aa} + a)$
>
> Surface area $=$ Sum of the areas of the trapeziums forming the sides plus the areas of the top and base of the frustum

**EXAMPLE 9.18**

A casting has a length of 2 m and its cross-section is a regular hexagon. The casting tapers uniformly along its length, the hexagon having a side of 200 mm at one end and 100 mm at the other. Calculate the volume of the casting.

In Fig. 9.17 the area of the hexagon of 200 mm side

$$= 6 \times \text{Area} \triangle ABO$$

In $\triangle AOC$, $\angle AOC = 30°$, $AC = 100$ mm

$$\therefore OC = \frac{AC}{\tan 30°} = \frac{100}{\tan 30°}$$

$$= 173 \text{ mm}$$

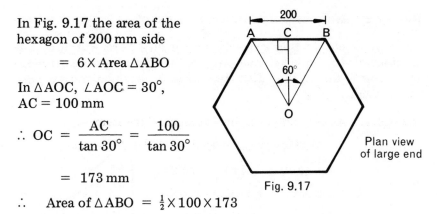

200

A  C  B

60°

O

Plan view of large end

Fig. 9.17

$$\therefore \quad \text{Area of } \triangle ABO = \tfrac{1}{2} \times 100 \times 173$$

$$\therefore \quad \text{Area of hexagon } = 6 \times \tfrac{1}{2} \times 100 \times 173 = 51\,900 \text{ mm}^2$$

The area of the hexagon of 100 mm side can be found in the same way. It is $13\,000 \text{ mm}^2$.

The casting is a frustum of a pyramid with $A = 51\,900$, $a = 13\,000$ and $h = 2000$

$$\therefore \text{Volume} = \tfrac{1}{3}h(A + \sqrt{Aa} + a)$$

$$= \tfrac{1}{3} \times 2000 \times (51\,900 + \sqrt{51\,900 \times 13\,000} + 13\,000)$$

$$= 6.06 \times 10^7 \text{ mm}^3$$

## The Ellipse

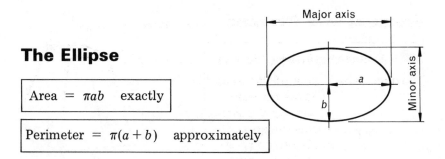

Major axis

Minor axis

a

b

| Area $= \pi a b$   exactly |

| Perimeter $= \pi(a + b)$   approximately |

### EXAMPLE 9.19

Find the area and the approxomate perimeter of an ellipse with a major axis of 200 mm and a minor axis of 150 mm.

$$\text{Area } = \pi a b = \pi \times 100 \times 75 = 23\,600 \text{ mm}^2$$

$$\text{Perimeter } = \pi(a + b) = \pi \times (100 + 75) = 550 \text{ mm approximately}$$

**EXAMPLE 9.20**

Part of an air ducting system calls for an ellipse whose area is $20\,000\,\text{mm}^2$ to be placed on a sheet metal cylinder whose diameter is $150\,\text{mm}$ (Fig. 9.18). Find the angle $\theta$ at which the cylinder must be cut.

We have     $\pi ab = 20\,000$

and since     $b = 75\,\text{mm}$

then     $a = \dfrac{20\,000}{\pi \times 75} = 84.9\,\text{mm}$

From Fig. 10.24,

$$\cos\theta = \frac{150}{2 \times 84.9} = 0.883$$

$$\theta = 28°$$

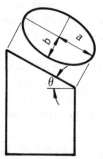

Fig. 9.18

# Exercise 9.2

**1)** A beam has a cross-sectional area of $18\,000\,\text{mm}^2$ and it is 6 m long. Calculate, in cubic metres, the volume of material in the beam.

**2)** A block of lead $0.15\,\text{m}$ by $0.1\,\text{m}$ by $0.075\,\text{m}$ is hammered out to make a square sheet $10\,\text{mm}$ thick. What are the dimensions of the square?

**3)** Calculate the volume of metal in a pipe which has a bore of $50\,\text{mm}$, a thickness of $8\,\text{mm}$ and a length of 6 m.

**4)** The vertical section of a cutting for a road is a trapezium $100\,\text{m}$ wide at the top and $30\,\text{m}$ wide at the bottom. If it is $15\,\text{m}$ deep and $40\,\text{m}$ long, calculate its volume.

**5)** The cross-section of a retaining wall is a trapezium as shown in Fig. 9.19. Its length is 8 m. Calculate the volume of the wall.

**6)** A hot-water cylinder whose length is $1.12\,\text{m}$ is to hold $200$ litres. Find the diameter of the cylinder in millimetres.

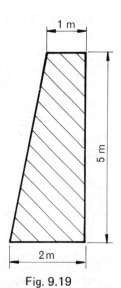

Fig. 9.19

**7)** Calculate the heating surface (in square metres) of a steam pipe whose external diameter is 60 mm and whose length is 8 m.

**8)** A small cone has a maximum diameter of 70 mm and a vertical height of 100 mm. Calculate is volume and its total surface area.

**9)** A cement silo is in the form of a frustum of a cone. It is 3 m high, 2.4 m diameter at the top and 1.2 m diameter at the bottom. Calculate the volume of cement that it will hold.

**10)** A metal bucket is 400 mm deep. It is 300 mm diameter at the top and 200 mm diameter at the bottom. Calculate the number of litres of water that the bucket will hold assuming that it is a frustum of a cone.

**11)** It is required to replace two pipes with bores of 28 mm and 70 mm respectively with a single pipe which has the same area of flow. Find the bore of this single pipe.

**12)** A pyramid has a square base of side 2 m and a height of 4 m. Calculate its volume and its total surface area.

**13)** A column is a regular octagon (8-sided polygon) in cross-section. It is 460 mm across flats at the base and it tapers uniformly to 300 mm across flats at the top. If it is 3.6 m high calculate the volume of material required to make it.

**14)** A bucket used on a crane is in the form of a frustum of a pyramid. Its base is a square of 600 mm side and its top is a square of 750 mm side. It has a depth of 800 mm.
(a) Calculate, in cubic metres, the volume of cement that it will hold.
(b) Twenty of these buckets of cement are emptied into a cylindrical cavity 4 m diameter. Calculate the depth to which the cavity will be filled.

**15)** A small turret roof is in the form of a pyramid. The base is a regular pentagon (5-sided polygon) with each side 3 m long. Its height is 4 m.
(a) Calculate the volume enclosed by the roof.
(b) Determine the total area of the inclined surfaces.

**16)** The ball of a float valve is a sphere 200 mm diameter. Calculate the volume and surface area of the ball.

**17)**  A hemispherical dome has a diameter of 7 m.

(a)  Calculate the volume that the dome encloses.

(b)  If the outside of the dome is to be painted, calculate the area, in square metres, which must be covered.

**18)**  A hemispherical dome has a radius of 4 m. It has its top 2 m cut off to provide for a horizontal laylight.

(a)  Calculate the volume enclosed by the dome below the laylight.

(b)  Find the area of internal plastering required to cover the curved surface below the laylight.

**19)**  A dome is in the form of a cap of a sphere. The base radius of the cap is 6 m and the height of the dome is 5 m. Calculate the air space contained in the dome and the inner curved surface area.

**20)**  The ball of a float valve is a sphere 200 mm diameter. It is immersed to a depth of 80 mm in water. How many litres of water does it displace?

**21)**  Find the area and approximate perimeter of an ellipse which has a major axis of 40 mm and a minor axis of 30 mm.

**22)**  An air duct is 1.20 m in diameter. At one point it is transformed into an ellipse of equal area. If the minor axis is 0.8 m, what is the length of the major axis?

**23)**  A cylindrical vessel, made of sheet metal, is 200 mm in diameter. It is cut at an angle of 40° as shown in Fig. 9.20. Find the area and the perimeter of the ellipse so formed.

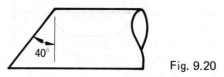

40°                    Fig. 9.20

**24)**  A swimming bath is 50 m long by 12 m wide. It is 3 m deep at the deep end and 1.5 m deep at the shallow end. Find the capacity of the bath in litres.

**25)**  A roof is in the form of a frustum of a rectangular pyramid. At the base the roof is 4 m long and 3 m wide whilst at the top it is 2 m long and 1.5 m wide. If it is 2 m high calculate the volume enclosed by the roof and the total surface area of the inclined faces.

**26)** A tapering column is 7 m tall and its cross-section is a regular octagon. The octagon is 3 m across flats at the base and 2 m across flats at the top.

(a) Calculate the volume of the column.

(b) If the material used in the construction of the column has a density of 2200 kg/m³, calculate the mass of the column.

**27)** The base of a ventilating turret is in the form of the frustum of a pyramid with a height of 1.3 m. The bottom of the turret is a square of side 2 m and the top is a square of side 1 m. Calculate the volume of the turret and the total area of the inclined surfaces.

**28)** The length of a conical nozzle is 600 mm long. The diameters at the ends are 150 mm and 220 mm respectively. Calculate the volume of water, in litres, that the nozzle will hold.

**29)** A lime kiln has the dimensions shown in Fig. 9.21. Calculate the volume enclosed by the kiln.

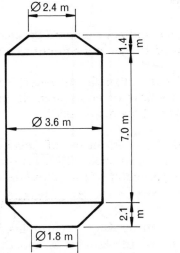

Fig. 9.21

**30)** A tub holding 58 litres of water when full is shaped like the frustum of a cone. If the radii at the ends of the tub are 400 mm and 300 mm, calculate the height of the tub.

**31)** A cupola is a cap of a 1.5 mm radius sphere. If it is 2 m high, calculate its volume and curved surface area.

**32)** A dome is in the form of the cap of a sphere. The radius at the base of the cap is 7 m and its height is 5 m. Calculate the area for plastering the inside of the dome.

# NUMERICAL METHODS FOR CALCULATING IRREGULAR AREAS AND VOLUMES

## Areas

An irregular area is one whose boundary does not follow a definite pattern, e.g. the cross-section of a river.

In these cases practical measurements are made and the results plotted to give a graphical display.

Various numerical methods may then be used to find the area.

## Mid-ordinate Rule

Suppose we wish to find the area shown in Fig. 9.22. Let us divide the area into a number of vertical strips, each of equal width $b$.

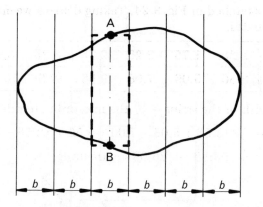

Fig. 9.22

Consider the 3rd strip, whose centre line is shown cutting the curved boundaries of the area at A and B respectively. Through A and B horizontal lines are drawn and these help to make the dotted rectangle shown. The rectangle has approximately the same area as that of the original 3rd strip, and this area will be $b \times AB$.

AB is called the mid-ordinate of the 3rd strip, as it is mid-way between the vertical sides of the strip.

To find the *whole area*, the areas of the other strips are found in a similar manner and then all are added together for the final result.

∴        Area = Width of strip × Sum of the mid-ordinates

A useful practical tip to avoid measuring each separate mid-ordinate is to use a strip of paper and mark off along its edge successive mid-ordinate lengths, as shown in Fig. 9.23. The total area will then be found by measuring the whole length marked out (in the case shown this is HR) and multiplying by the strip width $b$.

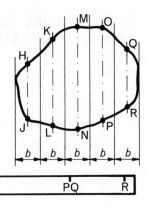

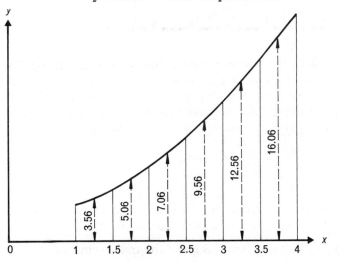

Fig. 9.23

### EXAMPLE 9.21

Find the area under the curve $y = x^2 + 2$ between $x = 1$ and $x = 4$

The curve is sketched in Fig. 9.24. Taking 6 strips we may calculate the mid-ordinates.

| $x$ | 1.25 | 1.75 | 2.25 | 2.75 | 3.25 | 3.75 |
|---|---|---|---|---|---|---|
| $y$ | 3.56 | 5.06 | 7.06 | 9.56 | 12.56 | 16.06 |

Since the width of the strips $= \frac{1}{2}$, the mid-ordinate rule gives

$$\text{Area} = \frac{1}{2} \times (3.56 + 5.06 + 7.06 + 9.56 + 12.56 + 16.06)$$

$$= \frac{1}{2} \times 53.86 = 26.93 \text{ square units}$$

Fig. 9.24

It so happens that in this example it is possible to calculate an exact answer. How this is done need not concern us at this stage, but by comparing the exact answer with that obtained by the mid-ordinate rule we can see the size of the error.

$$\text{Exact answer} = 27 \text{ square units}$$

$$\text{Approximate answer (using the mid-ordinate rule)}$$

$$= 26.93 \text{ square units}$$

$$\therefore \qquad\qquad \text{Error} = 0.07 \text{ square units}$$

$$\therefore \qquad \text{Percentage error} = \frac{0.07}{27} \times 100 = 0.26\%$$

From the above it is clear that the mid-ordinate rule gives a good approximation to the correct answer.

## Trapezoidal Rule

Consider the area having boundary ABCD shown in Fig. 9.25.

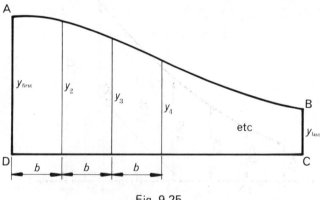

Fig. 9.25

The area is divided into a number of vertical strips of equal width $b$.

Each vertical strip is assumed to be a trapezium. Hence the third strip, for example, will have an area $= b \times \frac{1}{2}(y_3 + y_4)$.

But

Area ABCD = Sum of all the vertical strips

$$= b \times \tfrac{1}{2}(y_{\text{first}} + y_2) + b \times \tfrac{1}{2}(y_2 + y_3) + b \times \tfrac{1}{2}(y_3 + y_4) + \ldots$$

$$= b[\tfrac{1}{2}y_{\text{first}} + \tfrac{1}{2}y_2 + \tfrac{1}{2}y_2 + \tfrac{1}{2}y_3 + \ldots + \tfrac{1}{2}y_{\text{last}}]$$

$$= b[\tfrac{1}{2}(y_{\text{first}} + y_{\text{last}}) + y_2 + y_3 + y_4 \ldots]$$

= Width of strips × [$\tfrac{1}{2}$(Sum of the first and last ordinates) + (Sum of the remaining ordinates)]

The accuracy of the trapezoidal rule is similar to that of the mid-ordinate rule. A comparison may be made by solving Example 9.21 using the trapezoidal rule as in Example 9.22.

### EXAMPLE 9.22

Find the area under the curve $y = x^2 + 2$ between $x = 1$ and $x = 4$

The curve is sketched in Fig. 9.26, the lengths of the ordinates having been calculated.

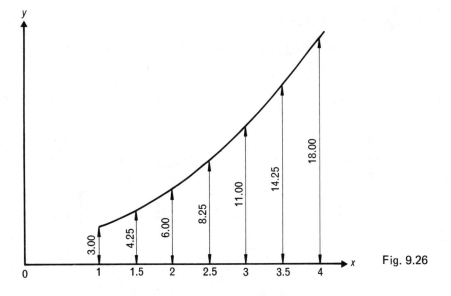

Fig. 9.26

Since the width of the strips $= \tfrac{1}{2}$, the trapezoidal rule gives

$$\text{Area} = \tfrac{1}{2} \times [\tfrac{1}{2}(3 + 18) + 4.25 + 6 + 8.25 + 11 + 14.25]$$

$$= \tfrac{1}{2} \times [10.5 + 43.75]$$

$$= 27.13 \text{ square units}$$

The exact answer is 27 square units and therefore

$$\text{Percentage error} = \frac{27.13 - 27}{27} \times 100 = 0.48\%$$

### EXAMPLE 9.23

The table gives the values of a force required to pull a trolley when measured at various distances from a fixed point in the direction of the force.

| $F$ (N) | 51 | 49 | 45 | 37 | 26 | 15 | 10 |
|---|---|---|---|---|---|---|---|
| $s$ (m) | 0 | 1 | 2 | 3 | 4 | 5 | 6 |

Calculate the total work done by this force.

The force–distance graph is plotted as shown in Fig. 9.27. The required work done is given by the shaded area under the curve. This may be found by dividing the area into strips and using the trapezoidal rule.

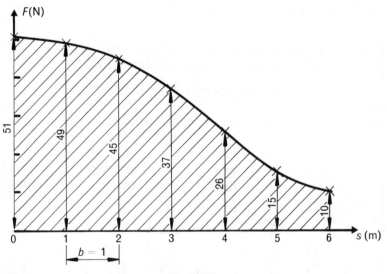

Fig. 9.27

Therefore

$$\text{Area} = 1[\tfrac{1}{2}(51 + 10) + (49 + 45 + 37 + 26 + 15)] = 203$$

Thus

Work done $= 203\,\text{Nm} = 203\,\text{J}$   (since joule = newton × metre)

# Simpson's Rule

The required area is divided into an *even* number of vertical strips of equal width $b$.

Then Simpson's rule gives

$$\text{Area} = \frac{b}{3}[(\text{Sum of the first and last ordinates})$$
$$+ \ 2(\text{Sum of the remaining odd ordinates})$$
$$+ \ 4(\text{Sum of the even ordinates})]$$

This rule usually gives a more accurate result than either the mid-ordinate or trapezoidal rules, but is slightly more complicated to use.

**Note.**    There must be an *even* number of strips.

### EXAMPLE 9.24

One boundary of a plot of land is a straight line 60 m long. The lengths of perpendicular offsets at 10 m intervals from this line to the curved boundary are as follows:

| Distance (m) | 0 | 10 | 20 | 30 | 40 | 50 | 60 |
|---|---|---|---|---|---|---|---|
| Length of offset (m) | 0 | 16 | 28 | 36 | 40 | 42 | 43 |

Draw a plan of the plot of land and find its area.

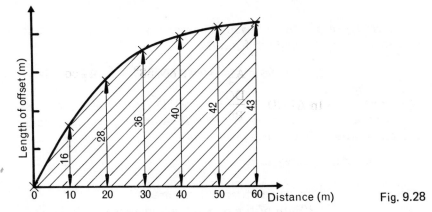

Fig. 9.28

The area under the graph gives the area of the plot and it is shown shaded in Fig. 9.28.

Simpson's rule gives

$$\text{Area} = \tfrac{10}{3}[(0+43) + 2(28+40) + 4(16+36+42)] = 1850$$

Therefore area of the plot of land is $1850\,\text{m}^2$.

# Volumes of Irregular Solids

All the methods explained in this chapter for finding irregular areas may be applied to finding volumes. The following example shows a typical problem solved by the use of Simpson's rule and the trapezoidal rule.

**EXAMPLE 9.25**

The diameters in metres of a felled tree trunk at one metre intervals along its length are as follows:

$$1.00, \ 0.90, \ 0.81, \ 0.74, \ 0.68, \ 0.64 \ \text{and} \ 0.61$$

Assuming that the cross-sections of the trunk are circular, estimate the volume of timber.

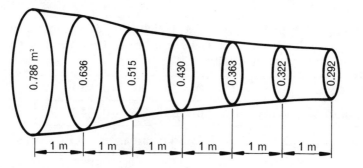

Fig. 9.29

The areas corresponding to each diameter are as shown in Fig. 9.29 and are calculated using the formula

$$\text{Area} = \frac{\pi}{4} \, (\text{Diameter})^2,$$

The graph of area against length could be plotted and the cross-sectional areas would be the lengths of ordinates. The area under this curve would then represent volume.

Using Simpson's rule we have

$$\text{Volume} = \frac{b}{3} \, [(\text{Sum of the first and last areas}) + 2(\text{Sum of the remaining odd areas}) + 4(\text{Sum of the even areas})]$$

$$= \tfrac{1}{3}[(0.786 + 0.292) + 2(0.515 + 0.363) + 4(0.636 + 0.430 + 0.322)]$$

$$= 2.79 \, \text{m}^3 \quad \text{the units being the result of multiplying} \\ \text{m}^2 \text{ (area) by m (length)}$$

An alternative layout using a table to show the calculations is often used:

| Area number | Area | Simpson's multiplier | Product |
|:---:|:---:|:---:|:---:|
| 1 | 0.786 | 1 | 0.786 |
| 2 | 0.636 | 4 | 2.544 |
| 3 | 0.515 | 2 | 1.030 |
| 4 | 0.430 | 4 | 1.720 |
| 5 | 0.363 | 2 | 1.726 |
| 6 | 0.322 | 4 | 1.288 |
| 7 | 0.292 | 1 | 0.292 |
| | | ∴ Total product = 8.386 | |

Hence     Volume = $\frac{1}{3}(8.386)$ = $2.795\,\text{m}^3$

Using the trapezoidal rule we have

Volume = $b[\frac{1}{2}$ (Sum of the first and last areas)
            + (Sum of the remaining areas)]

   = $1[\frac{1}{2}(0.786 + 0.292)$
         $+ 0.636 + 0.515 + 0.430 + 0.363 + 0.322)]$

   = $2.805\,\text{m}^3$

These results are reasonably close and we could safely assume that the volume of timber is 2.8 cubic metres.

# Prismoidal Rule for Calculating Volumes

The prismoidal formula is a general formula by which the volume of any prism, pyramid or frustum of a pyramid may be found. It is

$$V = \frac{h}{6}(A_1 + 4A_m + A_2)$$

where

$A_1$ = Area of one end of object,   $A_2$ = Area of other end

$A_m$ = Area of the section mid-way between the two end surfaces

$h$ = Distance between areas $A_1$ and $A_2$

It will be noticed that the prismoidal rule is really Simpson's rule for two strips.

**EXAMPLE 9.26**

A cutting is 180 m long. The cross-sectional areas at the ends of the cutting are 22 m$^2$ and 25 m$^2$ whilst the cross-sectional area at the middle of the cutting is 28 m$^2$. Find the volume of earth excavated.

We have   $h = 180$,   $A_1 = 22$,   $A_m = 28$   and   $A_2 = 25$

$$\text{Volume of earth excavated} = \frac{180}{6}(22 + 4 \times 28 + 25) = 4770\,\text{m}^3$$

# Exercise 9.3

It is suggested that the following questions are solved using at least two of the methods covered in the preceding text, i.e., using the trapezoidal rule, using the mid-ordinate rule, or by using Simpson's rule.

1) The table below gives corresponding values of $x$ and $y$. Plot the graph and by using the mid-ordinate rule find the area under the graph.

| $x$ | 1.5 | 1.7 | 1.9 | 2.1 | 2.3 | 2.5 | 2.7 | 2.9 | 3.1 |
|-----|-----|-----|-----|-----|-----|-----|-----|-----|-----|
| $y$ | 800 | 730 | 622 | 528 | 438 | 366 | 306 | 262 | 214 |

2) The table below gives corresponding values of two quantities $A$ and $x$. Draw the graph and hence find the area under it. (Plot $x$ horizontally.)

| $A$ | 53.2 | 35 | 22.2 | 21.8 | 24.2 | 23.6 | 18.7 | 0 |
|-----|------|-----|------|------|------|------|------|---|
| $x$ | 0 | 1 | 2 | 3 | 4 | 5 | 6 | 7 |

**3)** Plot the curve given by the following values of $x$ and $y$ and hence find the area included by the curve and the axes of $x$ and $y$.

| $x$ | 1 | 2 | 3 | 4 | 5 |
|---|---|---|---|---|---|
| $y$ | 1 | 0.25 | 0.11 | 0.063 | 0.040 |

**4)** Plot the curve of $y = 2x^2 + 7$ between $x = 2$ and $x = 5$ and find the area under this curve.

**5)** Plot the graph of $y = 2x^3 - 5$ between $x = 0$ and $x = 3$ and find the area under the curve.

**6)** A series of soundings taken across a section of a river channel are given in Fig. 9.30. Find an approximate value for the cross-sectional area of the river at this section.

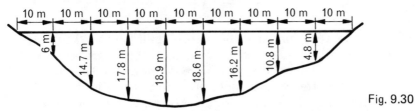

Fig. 9.30

**7)** The cross-sectional areas of a tree trunk are given in the table below. Find its volume.

| Distance from one end (m) | 0 | 1 | 2 | 3 | 4 | 5 | 6 |
|---|---|---|---|---|---|---|---|
| Area (m²) | 5.1 | 4.1 | 3.4 | 2.7 | 2.2 | 1.8 | 1.3 |

**8)** The width of a river at a certain section is 60 m. Soundings of the depth of the river taken at this section are recorded as follows:

| Distance from left bank (m) | 0 | 5 | 10 | 15 | 20 | 25 | 30 | 35 | 40 | 45 | 50 | 55 | 60 |
|---|---|---|---|---|---|---|---|---|---|---|---|---|---|
| Sounding depth (m) | 4 | 8 | 9 | 19 | 30 | 35 | 30 | 24 | 20 | 16 | 10 | 8 | 0 |

Plot the above information and drawing a fair curve through the points, calculate the cross-sectional area of the river at this section in m². If the volume of water flowing past this point per second is 10000 m³, calculate the speed, in m/s, at which the river is flowing.

*Hint:* Volume of flow per second = (Cross-sectional area)
$\qquad\qquad\qquad\qquad\qquad$ × (Velocity of flow)

**9)** Observation by surveyors show that the cross-sectional areas at 100 m intervals of a cutting are as shown in Fig. 9.31. Find the volume of soil required to fill the cutting.

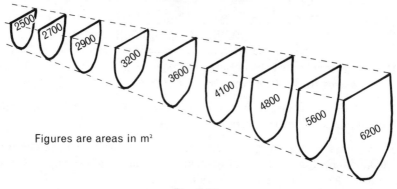

Figures are areas in m²

Fig. 9.31

**10)** A drain 15 m long is to be laid on level ground. The bottom of the trench is to be 0.5 m wide over the entire length. The depths at the ends of the trench are 0.7 m and 0.9 m whilst half way along the trench the depth is 0.85 m. The width of the top of the trench at the shallow end is 0.62 m, at the middle section it is 0.66 m and at the deep end it is 0.59 m. Calculate the amount of earth excavated.

**11)** A cutting for a road is 150 m long. The cross-sectional areas at the two ends and at the middle are 24 m², 31 m² and 29 m². Calculate the amount of earth excavated.

**12)** Part of a ventilation system consists of an equal tapered elliptical duct. It is 300 mm by 240 mm at one end and 220 mm by 160 mm at the other end. If it is 600 mm long, use the prismoidal rule to calculate the volume of the duct.

# 10. VOLUMES AND SURFACE AREAS
## (using THEOREMS OF PAPPUS)

*After reaching the end of this chapter you should be able to:*

1. *Use the theorem of Pappus to calculate the volume of a solid.*

2. *Use the theorem of Pappus to calculate the surface area of a solid.*

## CENTROIDS OF AREAS

The centroid of an area is at the point which corresponds to the centre of gravity of a lamina of the same shape as the area. A thin flat sheet of metal of uniform thickness is an example of a lamina. For calculation purposes the centroid of an area is the point at which the total area may be considered to be situated.

The positions of centroids of the areas met most frequently in mensuration problems are given in Fig. 10.1.

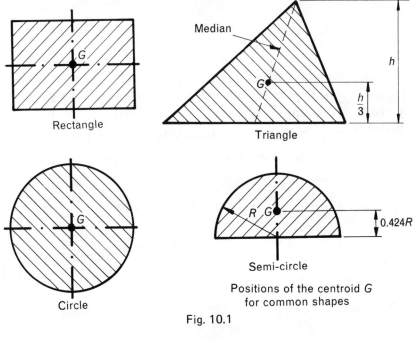

Positions of the centroid *G*
for common shapes

Fig. 10.1

# THEOREMS OF PAPPUS (or GULDINUS)

## Theorem 1

If a plane area rotates about a line in its plane (which does not cut the area) then the volume generated is given by the equation

**Volume = Area × Length of path of its centroid**

**EXAMPLE 10.1**

Find the volume of the circular ring of concrete shown in Fig. 10.2.

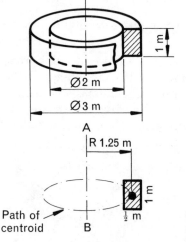

Fig. 10.2

Consider that the volume of the ring is made by one revolution of the cross-sectional area shown about the axis AB. Pappus' theorem 1 states

Volume = Area × Length of path of centroid of area

Hence, for one revolution of the area about AB, we have

$$\text{Volume} = (1 \times \tfrac{1}{2}) \times 2\pi\,1.25 = 1.25\theta = 3.93 \text{ m}^3$$

This can be checked by considering the ring as shown in Fig. 10.3.

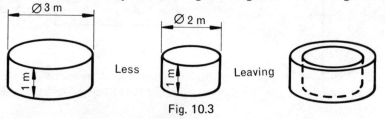

Fig. 10.3

Vol. of ring $= $ (Vol. of 3 m diam. disc) $-$ (Vol. of 2 m diam. disc)

$$= \left(\frac{\pi 3^2}{4}\right) \times 1 - \left(\frac{\pi 2^2}{4}\right) \times 1 \ = \ 3.93 \, \text{m}^3$$

## COMPOSITE AREAS

A composite area is an area which comprises two or more simple shapes as for example in Fig. 10.4 where the cross-sectional area comprises a rectangle and a triangle.

When using the theorem of Pappus with a composite area rotate each simple shape individually and then add together the separate volumes generated.

### EXAMPLE 10.2

Calculate the volume of concrete required to make the retaining wall shown in Fig. 10.4.

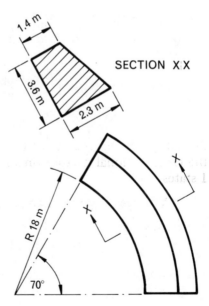

Fig. 10.4

Fig. 10.5 shows the given sectional area as comprising a rectangle and a triangle and the distances of their centroids from the axis of rotation.

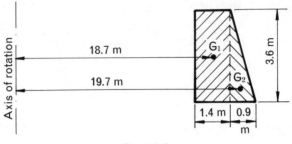

Fig. 10.5

Now

$$\text{Volume of wall} = \begin{bmatrix} \text{Volume generated} \\ \text{by the revolution,} \\ \text{through } 70°, \text{ of the} \\ \text{rectangular area} \end{bmatrix} + \begin{bmatrix} \text{Volume generated} \\ \text{by the revolution,} \\ \text{through } 70°, \text{ of the} \\ \text{triangular area} \end{bmatrix}$$

But theorem 1 of Pappus gives:

$$\text{Volume} = (\text{Area}) \times (\text{Length of path of centroid})$$

and if this is used to find each of the generated volumes then we have

$$\text{Volume of wall} = \left[ \left( \begin{matrix} \text{Rectangular} \\ \text{area} \end{matrix} \right) \times \left( \begin{matrix} \text{Length of path} \\ \text{of centroid} \end{matrix} \right) \right]$$

$$+ \left[ \left( \begin{matrix} \text{Triangular} \\ \text{area} \end{matrix} \right) \times \left( \begin{matrix} \text{Length of path} \\ \text{of centroid} \end{matrix} \right) \right]$$

$$= \left[ \left( 3.6 \times 1.4 \right) \times \left( \frac{70}{360} \times 2 \times \pi \times 18.7 \right) \right]$$

$$+ \left[ \left( \frac{1}{2} \times 0.9 \times 3.6 \right) \times \left( \frac{70}{360} \times 2 \times \pi \times 19.7 \right) \right]$$

$$= 115 + 39 = 154 \, \text{m}^3$$

**EXAMPLE 10.3**

A uniform solid circular cylinder of diameter 60 mm and height 15 mm is to be made into a pulley wheel by cutting a groove round the curved surface. The cross-section of the groove is to be a semi-circle of diameter 10 mm.

The distance of the centroid of the semi-circular cross-sectional area of the groove from its diameter is $0.424 \times \dfrac{10}{2} = 2.12\,\text{mm}$, so the distance of the centroid from the axis of the cylinder

$$= \frac{60}{2} - 2.12 = 27.88\,\text{mm}$$

(see Fig. 10.6).

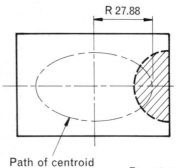

R 27.88

Path of centroid

Fig. 10.6

By Pappus' theorem 1:

Volume of cut-away portion = Area of groove
$\times$ Length of path of centroid

and so for one revolution of the area about the cylinder axis,

Volume of cut-away portion $= \frac{1}{2}\pi 5^2 \times 2\pi(27.88) = 6880\,\text{mm}^3$

but

Volume of pulley = Volume of cylinder − Volume of cut-away

$$= \frac{\pi}{4}(60)^2(15) - 6880 = 35\,500\,\text{mm}^3$$

## Theorem 2

If an arc rotates about a line in its plane (which does not cut the arc) then the surface area generated is given by the equation

**Area = Length of arc × Length of path of the arc centroid**

**EXAMPLE 10.4**

Find the surface area of the circular ring of concrete shown in Fig. 10.7.

Consider that the surface area of the ring is made by one revolution of a wire, bent to the shape of the perimeter of the rectangular cross-section around the polar axis of the ring (Fig. 10.7).

Pappus' theorem 2 states that

Area swept = Length of arc × Path of centroid

Therefore

Ring surface area $= (1 + 1 + \frac{1}{2} + \frac{1}{2}) \times (2 \times 1.25 \times \pi) = 23.5\,\text{m}^2$

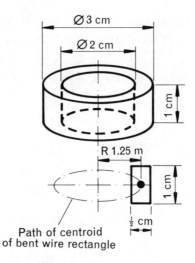

Path of centroid
of bent wire rectangle

Fig. 10.7

# Exercise 10.1

**1)** A reinforced concrete ring 4 m internal diameter whose cross-section is a square of 0.4 m side is to be cast. Calculate the volume of concrete required. If the ring is to be painted, calculate the area to be covered.

**2)** The ramp (Fig. 10.8) is to surround a monument. Calculate the volume of material needed to make the ramp.

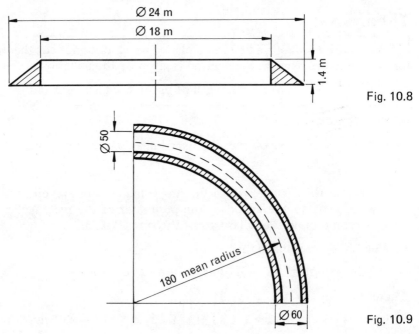

Fig. 10.8

Fig. 10.9

3) The right-angled bend shown in Fig. 10.9 is to be formed on a copper pipe. Find the volume of the bend.

4) Part of the ventilation system consists of the ducting shown in Fig. 10.10. Calculate the total outer curved surface area of the ducting which is required for estimating the cost of decorating it.

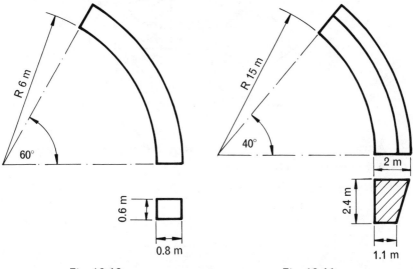

Fig. 10.10                        Fig. 10.11

5) Fig. 10.11 shows a retaining wall. Calculate the volume of concrete required to make it.

6) A concrete channel (Fig. 10.12) is to be made. What volume of concrete is required?

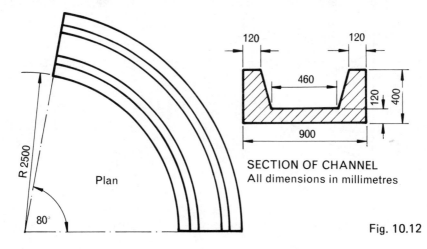

Plan

SECTION OF CHANNEL
All dimensions in millimetres

Fig. 10.12

**7)** A metal casting (Fig. 10.13) is to be made. Find the volume of metal required.

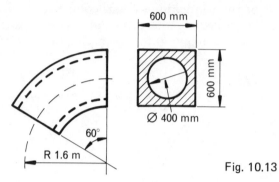

Fig. 10.13

**8)** A triangular ring (Fig. 10.14) is required as part of a design.

(a) What is the total surface area of the ring?

(b) Calculate the volume of the ring.

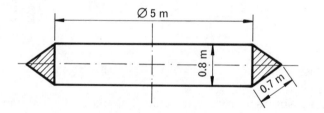

Fig. 10.14

**9)** Find the volume of the turned component shown in Fig. 10.15:

(a) using a theorem of Pappus,

(b) by considering the shape as the frustum of a cone.

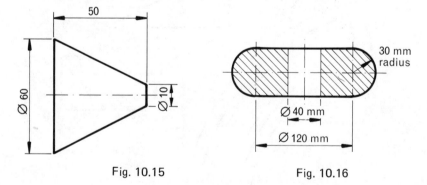

Fig. 10.15                Fig. 10.16

**10)** Find the volume of the ring shown in Fig. 10.16.

# 11. PRACTICAL TRIGONOMETRY

After reaching the end of this chapter you should be able to:

1. Apply trigonometry to solutions of practical problems.
2. Calculate the area of any triangle using the formulae $\sqrt{s(s-a)(s-b)(s-c)}$ and $\frac{1}{2}ab\sin\theta$.
3. Solve problems on triangles and quadrilaterals involving the formulae for the areas of triangles.

## SOLVING PRACTICAL BUILDING PROBLEMS

Trigonometry is frequently used to solve problems which arise in practice. Some of these problems and their solutions are shown in the examples which follow.

### EXAMPLE 11.1

The rise of a pitched roof is 1.6 metres and the roof makes an angle of $26°$ with the horizontal. Calculate the length of the common rafters.

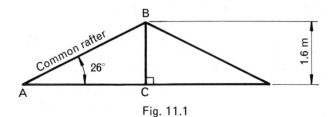

Fig. 11.1

The conditions are shown in Fig. 11.1, the common rafter being AB. Now in $\triangle ABC$,

$$\frac{AB}{BC} = \operatorname{cosec} 26°$$

$$AB = BC \operatorname{cosec} 26° = 1.6 \operatorname{cosec} 26° = 3.65$$

Hence the common rafters are 3.65 metres long.

**EXAMPLE 11.2**

Fig. 11.2 shows a roof truss. Calculate the lengths of the members BC, BD and AC.

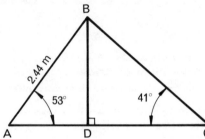

Fig. 11.2

In $\triangle ABD$,

$$\frac{BD}{AB} = \sin 53°$$

$\therefore$        $BD = AB \sin 53° = 2.44 \sin 53° = 1.95$

In $\triangle BCD$,

$$\frac{BC}{BD} = \operatorname{cosec} 41°$$

$\therefore$        $BC = BD \operatorname{cosec} 41° = 1.95 \operatorname{cosec} 41° = 2.97$

To find the length of AC we must first find the lengths of AD and DC.

In $\triangle ABD$,    $\dfrac{AD}{AB} = \cos 53°$

$\therefore$        $AD = AB \cos 53° = 2.44 \cos 53° = 1.47$

In $\triangle BCD$,    $\dfrac{DC}{BC} = \cos 41°$

$\therefore$        $DC = BC \cos 41° = 2.97 \cos 41° = 2.24$

Thus        $AC = AD + DC = 1.47 + 2.24 = 3.71$

Hence BD is 1.95 m. BC is 2.97 m and AC is 3.71 m.

# ANGLE OF ELEVATION

If you look upwards at an object the angle formed between the horizontal and your line of sight is called the *angle of elevation* (Fig. 11.3).

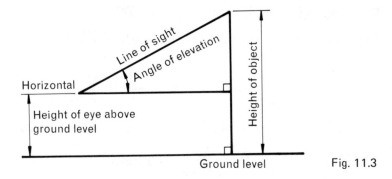

Horizontal

Height of eye above ground level

Line of sight

Angle of elevation

Height of object

Ground level

Fig. 11.3

## EXAMPLE 11.3

To find the height of a tower a surveyor sets up his theodolite 100 m from the base of the tower. He finds the angle of elevation of the top of the tower to be 30°. If the instrument is 1.5 m from the ground, what is the height of the tower?

In Fig. 11.4,

$$\frac{BC}{AB} = \tan 30°$$

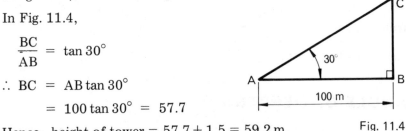

$$\therefore BC = AB \tan 30°$$

$$= 100 \tan 30° = 57.7$$

Hence, height of tower = 57.7 + 1.5 = 59.2 m.

Fig. 11.4

## EXAMPLE 11.4

To find the height of a pylon, a surveyor sets up a theodolite some distance from the base of the pylon and finds that the angle of elevation to the top of the pylon to be 30°. He then moves 60 m nearer to the pylon and finds that the angle of elevation is 42°. Find the height of the pylon assuming that the ground is horizontal and that the theodolite stands 1.5 m above the ground.

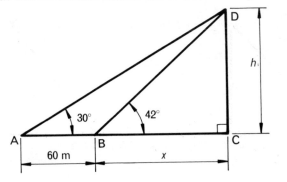

Fig. 11.5

Referring to Fig. 11.5, let $BC = x$ and $DC = h$.

In $\triangle ACD$,                                    In $\triangle BDC$,

$$\frac{DC}{AC} = \tan 30°$$                          $$\frac{DC}{BC} = \tan 42°$$

$\therefore$    $DC = AC \tan 30°$              $\therefore$    $DC = BC \tan 42°$

or    $h = 0.5774(x + 60)$  [1]        or    $h = 0.9004x$  [2]

From equation [2], $x = \dfrac{h}{0.9004} = 1.1106h$

Substituting for $x$ in equation [1] gives

$$h = 0.5774(1.1106h + 60) = 0.6413h + 34.64$$

$$h - 0.6413h = 34.64$$

or                $0.3587h = 34.64$

from which                $h = \dfrac{34.64}{0.3587} = 96.6 \text{ m}$

Hence the height of the pylon is $96.6 + 1.5 = 98.1 \text{ m}$.

# ANGLE OF DEPRESSION

If you look down at an object, the angle formed between the horizontal and your line of sight is called the *angle of depression* (Fig. 11.6).

### EXAMPLE 11.5

From the top floor window of a house, 14 m above ground level, the angle of depression of an object in the street is $52°$. How far is the object from the house?

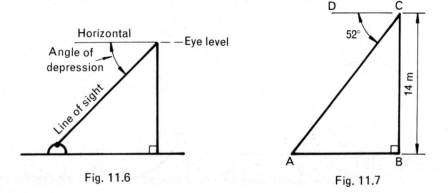

Fig. 11.6                                    Fig. 11.7

The conditions are shown in Fig. 11.7. Since the angle of depression is $52°$, $\angle ACD = 52°$.

$$\angle ACB = 90° - 52° = 38°$$

In $\triangle ABC$, $\quad \dfrac{AB}{CB} = \tan 38°$

$\therefore \qquad\qquad AB = CB \tan 38° = 14 \tan 38° = 10.9$

Hence the object is $10.9\,$m from the house.

# BEARINGS

The four cardinal directions are North, South, East and West (Fig. 11.8). The directions NE, NW, SE and SW are frequently used and are as shown in the diagram. A bearing of N20°E means an angle $20°$ measured from N towards E as shown in Fig. 11.9. Similarly a bearing of S40°E means an angle of $40°$ measured from S towards E (Fig. 11.10). A bearing of N50°W means an angle of $50°$ measured from N towards W (Fig. 11.11).

*Bearings quoted in this way are always measured from N and S and never from E and W.*

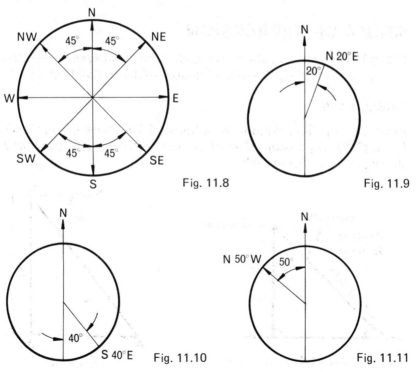

Fig. 11.8

Fig. 11.9

Fig. 11.10

Fig. 11.11

Another way of stating bearings is to measure the angle from N in a clockwise direction, N being taken as 0°. Three figures are always stated. For example 005° is written instead of 5° and 035° instead of 35° and so on. E will be 090°, S 180° and W 270°. Some typical bearings are shown in Fig. 11.12.

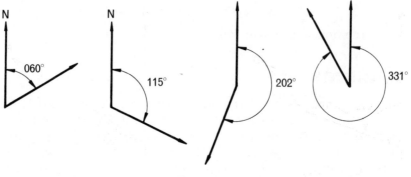

Fig. 11.12

## EXAMPLE 11.6

In making a survey it is found that B is a point due east of a point A and a point C is 6 km due south of A. The distance BC is 7 km. Calculate the bearing of C from B.

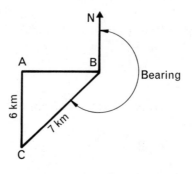

Fig. 11.13

In right-angled $\triangle ABC$, Fig. 11.13 we have

$$\sin \angle B = \frac{AC}{BC} = \frac{6}{7}$$

$$\therefore \qquad \angle B = 59°$$

The bearing of C from B $= 270° - 59° = 211°$

**EXAMPLE 11.7**

Fig. 11.14 represents a survey of a plot of land. Calculate the distance PB and the bearing of B from P.

Draw the triangles PAC, ABD and BFP as shown in Fig. 11.15.

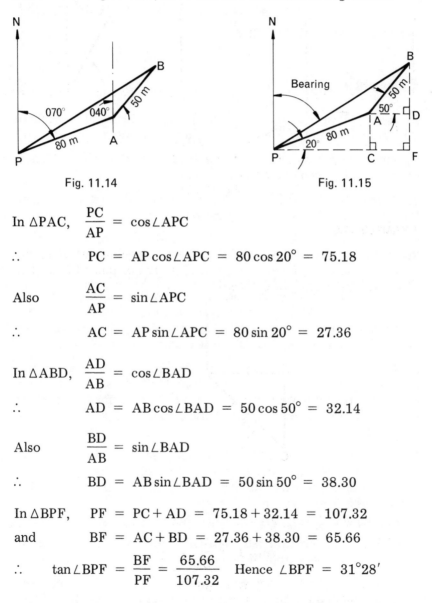

Fig. 11.14                    Fig. 11.15

In $\triangle$PAC,    $\dfrac{PC}{AP} = \cos\angle APC$

$\therefore$    $PC = AP\cos\angle APC = 80\cos 20° = 75.18$

Also    $\dfrac{AC}{AP} = \sin\angle APC$

$\therefore$    $AC = AP\sin\angle APC = 80\sin 20° = 27.36$

In $\triangle$ABD,    $\dfrac{AD}{AB} = \cos\angle BAD$

$\therefore$    $AD = AB\cos\angle BAD = 50\cos 50° = 32.14$

Also    $\dfrac{BD}{AB} = \sin\angle BAD$

$\therefore$    $BD = AB\sin\angle BAD = 50\sin 50° = 38.30$

In $\triangle$BPF,    $PF = PC + AD = 75.18 + 32.14 = 107.32$

and    $BF = AC + BD = 27.36 + 38.30 = 65.66$

$\therefore$    $\tan\angle BPF = \dfrac{BF}{PF} = \dfrac{65.66}{107.32}$    Hence $\angle BPF = 31°28'$

Thus the bearing of B from P is $90° - 31°28' = 58°32'$.

The distance PB may be found by using trigonometry or by using Pythagoras' theorem. Using trigonometry,

$$\frac{PB}{BF} = \operatorname{cosec} \angle BPF$$

∴          $PB = BF \operatorname{cosec} \angle BPF = 65.66 \operatorname{cosec} 31°28' = 125.8$

Hence the distance PB is 125.8 metres.

## Exercise 11.1

**1)** A straight stretch of road is 90 m long and rises at an angle of $20°$ to the horizontal. Find the vertical rise of this stretch of road.

**2)** The main rafters of a lean-to roof are each 3.96 m long. What is the span of the roof if the rise is 1.37 m?

**3)** A roof is pitched $31°$ to the horizontal and its rise is 2.62 m. Calculate the length of the main rafter.

**4)** A lean-to roof has a rise of 1.37 m and a span of 2.59 m. Find the angle that the roof makes with the horizontal.

**5)** A drain slopes downwards from a point A at an angle of $11°$ to the horizontal for 12 m to a point B. From B it slopes downwards at an angle of $14°$ from the horizontal for a further 13 m to a point C. From C it continues to D a distance of 25 m at an angle of $8°$ to the horizontal. Calculate the distance of D below A.

**6)** A road runs in a direction S30°E from a point A for a distance of 120 m to a point B. It then runs for a further distance of 140 m to a point C in a direction S35°E. Find: (a) how far C is south of A, (b) how far C is east of A, (c) the distance AC, (d) the bearing of C from A.

**7)** A tower is 25 m high. A man standing some distance from the tower finds the angle of elevation to the top of the tower to be $59°$. How far is the man standing from the foot of the tower, if his eye level is 1.5 m above ground level?

**8)** A man whose eye level is 1.5 m above ground level is 15 m away from a tower 20 m tall. Determine the angle of elevation of the top of the tower from his eyes.

**9)** A man standing on top of a mountain 1200 m high observes the angle of depression of the top of a steeple to be $43°$. If the height of the steeple is 50 m, how far is it from the mountain?

**10)** To find the height of a tower a surveyor stands some distance away from its base and he observes the angle of elevation to the top of the tower to be 45°. He then moves 80 m nearer to the tower and he then finds the angle of elevation to be 60°. Find the height of the tower. Assume the theodolite stands 1.5 m above ground level.

**11)** A tower is known to be 60 m high. A surveyor, using a theodolite, stands some distance away from the tower and measures the angle of elevation as 38°. How far away from the tower is he, if the theodolite stands 1.5 m above ground level? If the surveyor moves 80 m further away from the tower, what is now the angle of elevation of the tower?

**12)** A surveyor stands 100 m from the base of a tower on which an aerial stands. He measures the angles of elevation to the top and bottom of the aerial as 58° and 56°. Find the height of the aerial.

**13)** Triangle ABC (Fig. 11.16) represents the cross-section of a cutting. A bridge spanning the cutting is to be designed. Calculate the length of AC so that the design can be made.

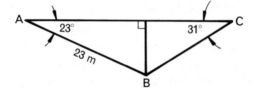

Fig. 11.16

**14)** Fig. 11.17 represents the framework for a footbridge AB = CD = EF = 2.4 m. Calculate the length of the members required to make this framework.

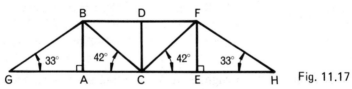

Fig. 11.17

**15)** To find the height of a feature a surveyor stands some distance away from its base and finds the angle of elevation to the top of the feature to be 49°. He then moves 60 m further away from the feature and finds the angle of elevation to be 38°. Calculate the height of the feature. Assume that the theodolite is 1.5 m above ground level.

**16)** A man standing on top of a building 50 m high is in line with two points A and B whose angles of depression are 17° and 21° respectively. Calculate the distance AB.

17) In surveying a plot of land it was found that A is 50 m due north of B. C is the point whose bearing from A is 150° and AC is 80 m. Calculate the distance BC and the bearing of C from B.

18) Three towns A, B and C lie on a straight road running east from A. B is 6 km from A and C is 22 km from A. Another town D lies to the north of this road and lies 10 km from both B and C. Calculate the distance of D from A and the bearing of D from A.

# SOLVING PRACTICAL ENGINEERING PROBLEMS

Some of the problems which occur in Mechanical Engineering are discussed in the sections which follow. In all of them trigonometry is used.

# CO-ORDINATE HOLE DIMENSIONS

In marking-out and in operating certain machine tools it is convenient to give the dimensions of holes relative to two axes which are at right angles to each other.

In graphical work the position of a point on a graph is specified by its co-ordinates (that is, the distances at which the point lies from the $x$- and $y$-axes respectively). Co-ordinate hole centres are specified in exactly the same way.

### EXAMPLE 11.8

Three holes are to have their centres equally spaced on a 50.00 mm pitch circle diameter as shown in Fig. 11.18. Calculate the co-ordinate dimensions of the holes, relative to the axes $Ox$ and $Oy$.

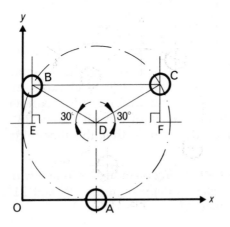

Fig. 11.18

**Hole A**
$x$ dimension $= 25.00$ mm
$y$ dimension $= 0$

To find the $x$ and $y$ dimensions for the holes B and C draw the $\triangle$DCF and the $\triangle$BED.

In $\triangle$BED,    $\dfrac{ED}{BD} = \cos 30°$

$\therefore$         $ED = BD(\cos 30°) = 25(\cos 30°) = 21.65\,\text{mm}$

and       $\dfrac{BE}{BD} = \sin 30°$

$\therefore$         $BE = BD(\sin 30°) = 25(\sin 30°) = 12.50\,\text{mm}$

Since $\triangle$BED is congruent with $\triangle$DCF,

    $ED = DF = 21.65\,\text{mm}$    and    $BE = CF = 12.50\,\text{mm}$

**Hole B**

    $x$ dimension $= 25.00 - ED = 25.00 - 21.65 = 3.35\,\text{mm}$

    $y$ dimension $= 25.00 + BE = 25.00 + 12.50 = 37.50\,\text{mm}$

**Hole C**

    $x$ dimension $= 25.00 + DF = 25.00 + 21.65 = 46.65\,\text{mm}$

    $y$ dimension $= 25.00 + CF = 25.00 + 12.50 = 37.50\,\text{mm}$

# Exercise 11.2

1) Fig. 11.19 shows 5 equally spaced holes on a 100 mm pitch circle diameter. Calculate their co-ordinate dimensions relative to the axes $Ox$ and $Oy$.

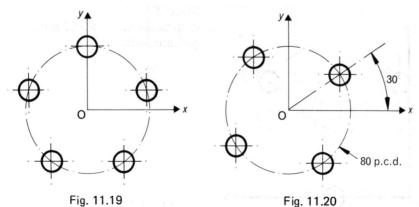

Fig. 11.19                         Fig. 11.20

2) 4 holes are equally spaced as shown in Fig. 11.20. Find their co-ordinate dimensions, relative to the axes O$x$ and O$y$.

3) Find the co-ordinate dimensions for the 3 holes shown in Fig. 11.21 relative to the axes O$x$ and O$y$. The holes lie on a 75 mm pitch circle diameter.

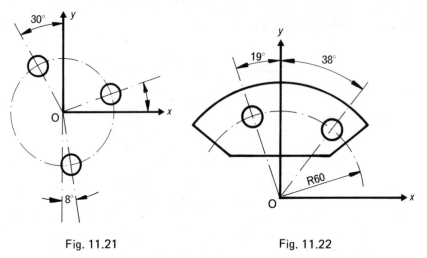

Fig. 11.21                    Fig. 11.22

4) Find the co-ordinate hole centres for the two holes shown in Fig. 11.22.

5) Find the co-ordinate hole dimensions for the two holes shown in Fig. 11.23.

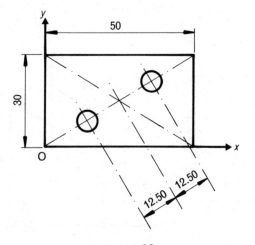

Fig. 11.23

# THE SINE BAR

A sine bar is used for measurements which require the angle to be determined closer than 5′. The method of using the instrument is shown in Fig. 11.24. The distance $l$ is usually made 100 mm or 200 mm to facilitate calculations. $h$ is the difference in height between the two rollers and $h = l \sin \theta$.

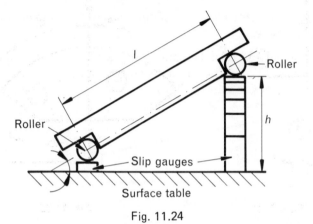

Fig. 11.24

### EXAMPLE 11.9

The angle of 15°42′ is to be checked on the metal block shown in Fig. 11.25. Find the difference in height between the two rollers which support the ends of the 200 mm sine bar.

Now

$$h = l \sin \theta$$

$$= 200 \times \sin 15°42′$$

$$= 200 \times 0.2706$$

$$= 54.12 \, \text{mm}$$

The difference in height of the slip gauges must therefore be 54.12 mm if the angle is correct.

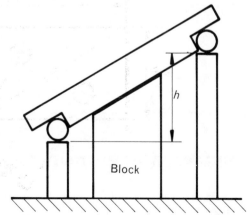

Fig. 11.25

## Exercise 11.3

1) Calculate the setting of a 100 mm sine bar to measure an angle of 27°15′.

2) Find the setting of a 200 mm sine bar to check a taper piece which has a taper of 1 in 10 on diameter. The piece is mounted in a similar way to that shown in Fig. 11.27.

3) Calculate the setting of a 200 mm sine bar to check a taper of 1 in 8 on diameter.

4) Find the setting of a 100 mm sine bar to check the taper piece shown in Fig. 11.26. The taper piece is mounted in a similar way to the component shown in Fig. 11.27.

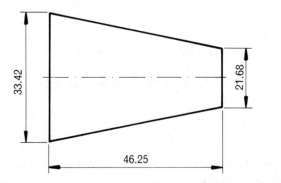

Fig. 11.26

5) Fig. 11.27 shows a component mounted on a 200 mm sine bar. Calculate the angle θ.

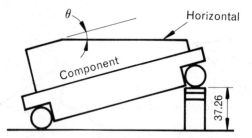

Fig. 11.27

## TRIGONOMETRY AND THE CIRCLE

In this section we shall consider the application of trigonometry and geometry to fine measurement but we must first look at some important geometrical theorems.

**(1)** If two circles are tangential to each other then the straight line which passes through the centres of the circles also passes through the point of tangency.

Thus the line AB joining the centres of the circles also passes through C, the point of tangency (Fig. 11.28) and AB is perpendicular to DE.

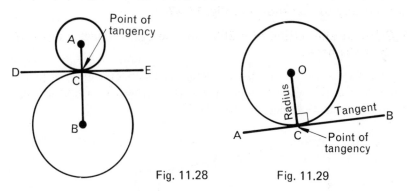

Fig. 11.28                Fig. 11.29

**(2)** If a line is tangential to a circle then it is at right angles to a radius drawn to the point of tangency.

Thus if AB is a tangent with C the point of tangency then the radius OC is at right angles to AB (Fig. 11.29).

**(3)** If from a point outside a circle tangents are drawn to the circle, then their lengths are equal.

Thus in Fig. 11.30 the lengths AC and BC are equal. It can also be proved that $\angle ACB$ is bisected by CO, O being the centre of the circle.

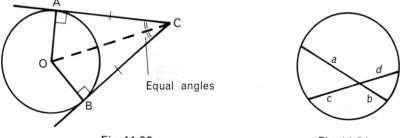

Fig. 11.30                Fig. 11.31

**(4)** If two chords intersect each other in a circle then the rectangle of the segments of the one equals the rectangle of the segments of the other.

Thus, in Fig. 11.31         $a \times b = c \times d$.

# REFERENCE ROLLERS AND BALLS

Sets of rollers can be obtained which are guaranteed to be within 0.002 mm for both diameter and roundness. By using rollers and balls many problems in measurement can be solved, some examples of which are given below.

**EXAMPLE 11.10**

A taper angle of $7°$ is to be checked by means of an adjustable gauge. The gauge is to be set by means of two rollers, one of 20 mm diameter and the other of 25 mm diameter. Find the centre distance $l$ between the rollers.

In Fig. 11.32, A and B are the centres of the rollers. E and D are the points where the rollers touch the top blade.

$$\angle AED = \angle BDE = 90°$$

(angles between a radius and a tangent).

Now draw AC parallel to ED,

In $\triangle CAB$,

$\angle CAB = 3°30'$   (half angle of taper)

$\angle ACB = 90°,$

$\quad BC = BD - AE = 12.5 - 10 = 2.50 \, mm$

$\quad AB = l$

$\therefore \dfrac{l}{2.50} = \text{cosec } 3°30'$

$\quad l = 2.50 \times \text{cosec } 3°30' = 40.95 \, mm$

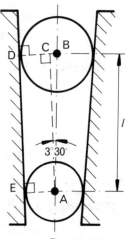

Fig. 11.32

**EXAMPLE 11.11**

A taper piece has a taper of 1 in 8 on the diameter. Two pairs of rollers 15.00 mm in diameter are used to check the taper as shown in Fig. 11.33. The measurement over the top rollers is 55.87 mm. Find:

a)  the measurement over the bottom rollers if the taper is correct,
b)  the bottom diameter of the job.

The first step is to find the angle of the taper. Using Fig. 11.34 we see that $\tan \alpha = \dfrac{0.5}{8}$   so   $\alpha = 3°34'$.

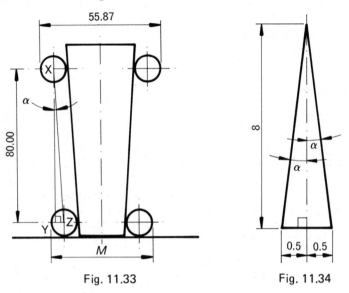

Fig. 11.33                    Fig. 11.34

a)  Referring to Fig. 11.33 we have in $\triangle XYZ$ that:

$$\angle \alpha = 3°34', \quad XY = 80.00 \, \text{mm}$$

but

$$\frac{YZ}{XY} = \tan \alpha$$

$\therefore$

$$YZ = 80 \times \tan 3°34' = 5.00$$

Thus

$$M = 55.87 - 2 \times 5.00 = 45.87 \, \text{mm}$$

b)  Referring to Fig. 11.35, we need to find AB.

$$\angle ABC = 90° - 3°34' = 86°26'$$

Since AB and BC are tangents, the line BZ bisects $\angle ABC$.
Hence           $ABZ = 43°13'$

In $\triangle ABZ$        $\dfrac{AB}{AZ} = \cot 43°13'$

$\therefore$

$$AB = AZ(\cot 43°13')$$

$$= 7.5(\cot 43°13') = 7.982 \, \text{mm}$$

$\therefore$  Bottom diameter $= M - (2 \times AB) - (2 \times \text{Radius of roller})$

$$= 45.87 - 7.982 - 2 \times 7.50 = 14.91 \, \text{mm}$$

**EXAMPLE 11.12**

The tapered hole shown was inspected by using two balls 25.00 and 20.00 mm diameter respectively. The measurements indicated in Fig. 11.36 were obtained. Find:

a) the included angle of taper $2\alpha$,
b) the top diameter $d$ of the hole.

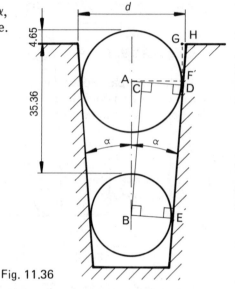

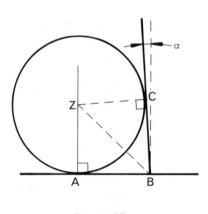

Fig. 11.35                Fig. 11.36

a) In the Fig. 11.35, A and B are the centres of the balls and E and D are points where the balls just touch the sides of the hole.

$\angle ADE = \angle BED = 90°$   (angles between radius and tangent)

Draw BC parallel to DE; then in $\triangle ABC$

$$AC = 12.50 - 10.00 = 2.50\,\text{mm}$$

$$AB = 35.36 + 4.65 - 12.50 + 10.00 = 37.51\,\text{mm}$$

Now   $\angle ACB = 90°$   and   $\angle ABC = \alpha$

∴      $\sin\alpha = \dfrac{AC}{AB} = \dfrac{2.50}{37.51}$   ∴   $\alpha = 3°49'$

∴      Included angle of taper $= 2\alpha = 2 \times 3°49' = 7°38'$.

b) To find $d$, draw AF horizontal and FG vertical.

Then                $d = 2 \times (AF + GH)$

In $\triangle AFD$,

$$AD = 12.50\,\text{mm}, \quad \angle FAD = \alpha, \quad \angle ADF = 90°$$

∴        $AF = AD \sec\alpha = 12.50 \times \sec 3°49' = 12.53\,\text{mm}$

In $\triangle$GFH,

GF $= 12.50 - 4.65 = 7.85\,\text{mm}$, $\angle$FGH $= 90°$, $\angle$GFH $= \alpha$,

$\therefore$    GH $= $ GF $\tan\alpha = 7.85 \times \tan 3°49' = 0.52\,\text{mm}$

Thus    $d = 2(\text{AF} + \text{GH}) = 2 \times (12.53 + 0.52) = 26.10\,\text{mm}$

## Exercise 11.4

1) A steel ball 40 mm in diameter is used to check the taper hole, a section of which is shown in Fig. 11.37. If the taper is correct, what is the dimension $x$?

2) A taper plug gauge is being checked by means of reference rollers and slip gauges. The set-up is as shown in Fig. 11.38. Find the included angle of the taper of the gauge and also the top and bottom diameters.

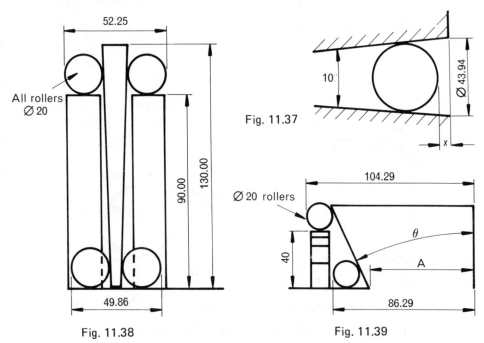

Fig. 11.37

Fig. 11.38

Fig. 11.39

3) Fig. 11.39 shows a dovetail being checked by rollers and slip gauges. Find the angle $\theta$ and the dimension $A$.

4) Calculate the dimension $M$ which is needed for checking the groove, a cross-section of which is shown in Fig. 11.40.

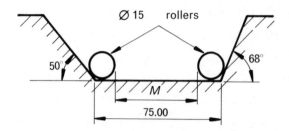

5) A tapered hole has a maximum diameter of 32.00 mm and an included angle of 16°. A ball having a diameter of 20 mm is placed in the hole. Calculate the distance between the top of the hole and the top of the ball.

6) Fig. 11.41 shows the dimensions obtained in checking a tapered hole. Find the included angle of taper of the hole and the top diameter $d$.

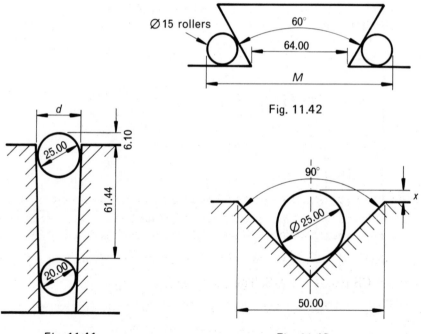

Fig. 11.42

Fig. 11.41

Fig. 11.43

7) Find the checking dimension $M$ for the symmetrical dovetail slide shown in Fig. 11.42

8) Fig. 11.43 shows a Vee block being checked by means of a reference roller. If the block is correct what is the dimension $x$?

# LENGTHS OF BELTS

There are two distinct cases, open belts and crossed belts, as shown in Fig. 11.44.

With the open belt, pulleys revolve in the *same* direction.

With the crossed belt, pulleys revolve in *opposite* directions.

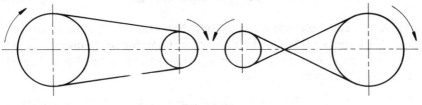

Fig. 11.44

## EXAMPLE 11.13

Find the length of an open belt which passes over two pulleys of 200 mm and 300 mm diameter respectively. The distance between the pulley centres is 900 mm.

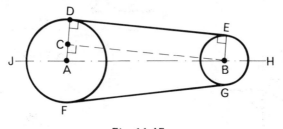

Fig. 11.45

Referring to Fig. 11.45, the total length of the belt is made up of the two straight lengths DE and FG and the arcs DJF and EHG.

$$\angle ADE = \angle BED = 90° \quad \text{(angle between radius and tangent)}$$

Draw CB parallel to DE. Then in $\triangle ABC$,

$$AC = AD - EB = 150 - 100 = 50 \text{ mm}$$

$$AB = 900 \text{ mm (given)}, \quad \angle ACB = 90°$$

$$\therefore \quad \cos C\hat{A}B = \frac{AC}{AB} = \frac{50}{900} = 0.0556 \quad \therefore \quad \angle C\hat{A}B = 86°49'$$

Now $\quad BC = AB \sin C\hat{A}B = 900 \times \sin 86°49' = 898.6 \text{ mm}$

Also $\qquad\qquad \angle EBH = \angle CAB = 86°49'$

Hence the arc EHG subtends an angle of $2 \times 86°49' = 173°38'$ at the centre.

$$\text{Length of arc EHG} = 2\pi \times 100 \times \frac{173°38'}{360°} = 303.0 \, \text{mm}$$

Now $\angle \text{DAJ} = 180° - \angle \text{CAB} = 180° - 86°49' = 93°11'$

The arc DJF therefore subtends an angle of $2 \times 93°11' = 186°22'$ at the centre.

$$\text{Length of arc DJF} = \frac{2\pi \times 150 \times 186°22'}{360°} = 488.0 \, \text{mm}$$

$\therefore$ Total length of belt $= 2 \times \text{BC} + \text{Arc EHG} + \text{Arc DJF}$

$$= 2 \times 898.6 + 303.0 + 488.0 = 2588 \, \text{mm}$$

**EXAMPLE 11.14**

Two pulleys 200 mm and 300 mm in diameter respectively are placed 1200 mm apart. They are connected by a closed belt. Find the length of the belt required.

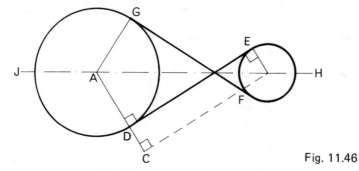

Fig. 11.46

The length of the belt is made up of two straight lengths ED and FG, arc GJD and arc EHF, as shown in Fig. 11.46.

$\angle \text{ADE} = \angle \text{DEB} = 90°$   (angle between radius and tangent)

Draw CB parallel to DE; then in $\triangle \text{ABC}$

$\text{AC} = 100 + 150 = 250 \, \text{mm}$,   $\text{AB} = 1200 \, \text{mm}$,   $\angle \text{ACB} = 90°$,

$\therefore \cos \hat{\text{CAB}} = \dfrac{\text{AC}}{\text{AB}} = \dfrac{250}{1200} = 0.2083$   $\therefore \angle \text{CAB} = 77°59'$

Thus $\text{BC} = \text{AB} \times \sin \hat{\text{CAB}} = 1200 \times \sin 77°59' = 1174 \, \text{mm}$

Also $\angle \text{EBA} = \angle \text{CAB} = 77°59'$

$\therefore \angle \text{EBH} = 180° - 77°59' = 102°1'$

The arc EHF therefore subtends an angle $2 \times 102°1'$ at the centre.

$$\therefore \quad \text{Length of arc EHF} \ = \ 2\pi \times 100 \times \frac{204°2'}{360°} \ = \ 356 \,\text{mm}$$

Similarly

$$\text{Length of arc GJD} \ = \ 2\pi \times 150 \times \frac{204°2'}{360°} \ = \ 534 \,\text{mm}$$

$$\therefore \quad \text{Total length of belt} \ = \ 2 \times 1174 + 356 + 534 \ = \ 3238 \,\text{mm}$$

## SCREW THREAD MEASUREMENT

When accurate screw thread measurement is required the method of 2 or 3 wire measurement is used. The 3-wire method is used when checking with a hand micrometer and the 2-wire method is used with bench micrometers. The methods are fundamentally the same and allow the pitch or effective diameters to be measured (Fig. 11.47).

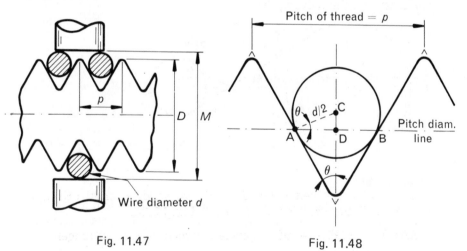

Fig. 11.47                         Fig. 11.48

For the most accurate results each type and size of thread requires wires of a certain size. The best wire size is one which just touches the thread flanks at the pitch diameter as shown in Fig. 11.48.

At the pitch line diameter the distance between the flanks of the thread is equal to half the pitch.

Hence in Fig. 11.48

$$\text{AB} \ = \ \frac{p}{2} \quad \text{and} \quad \text{AD} \ = \ \frac{p}{4}$$

In $\triangle ADC$,     $\dfrac{AC}{AD} = \sec \theta$

$\therefore$     $AC = AD \times \sec \theta$

$\therefore$     $\dfrac{d}{2} = \dfrac{p}{4} \sec \theta \quad \therefore \quad d = \dfrac{p}{2} \sec \theta$

For a metric thread,     $\theta = 30°$ and hence the best wire size is

$$d = \dfrac{p}{2} \sec 30° = 0.5774p$$

# Formula for Checking the Form of a Metric Thread

Although the best wire size should be used, in practice the wire used will vary a little from this best size. In order to determine the distance over the wires for a specific thread ($M$ in Fig. 11.47) the following formula is used

$$M = D - \dfrac{5p}{6} \cot \theta + d(\operatorname{cosec} \theta + 1)$$

For a metric thread, the formula becomes

$$M = D - \dfrac{5p}{6} \cot 30° + d(\operatorname{cosec} 30° + 1)$$

$$= D - 14434p + d(2 + 1) = D - 1.4434p + 3d$$

### EXAMPLE 11.15

A metric thread having a major diameter of 20 mm and a pitch of 2.5 mm is to be checked using the best wire size for this particular thread. Find:

a) the best wire size,
b) the measurement over the wires if the thread is correct.

a)  The best wire size is

$$d = 0.5774p = 0.5774 \times 2.5 = 1.4435 \text{ mm}$$

b)  The measurement over the wires is

$$M = D - 1.4434p + 3d$$

$$= 20 - 1.4434 \times 2.5 + 3 \times 1.4435 = 20.722 \text{ mm}$$

# Exercise 11.5

1) A belt passes over a pulley 1200 mm in diameter. The angle of contact between the pulley and the belt is 230°. Find the length of belt in contact with the pulley.

2) An open belt passes over two pulleys 900 mm and 600 mm in diameter respectively. If the centres of the pulleys are 1500 mm apart, find the length of the belt required.

3) Two pulleys of diameters 1400 mm and 900 mm respectively, with centres 4.5 m apart, are connected by an open belt. Find its length.

4) An open belt connects two pulleys of diameters 120 mm and 300 mm with centres 300 mm apart. Calculate the length of the belt.

5) A crossed belt passes over two pulleys each of 450 mm diameter. If their centres are 600 mm apart, calculate the length of the belt.

6) If, in Question 2, a crossed belt is used, what will be its length?

7) A crossed belt passes over two pulleys 900 mm and 1500 mm in diameter respectively, which have their centres 6 m apart. Find its length.

8) A crossed belt passes over two pulleys, one of 280 mm diameter and the other of 380 mm diameter. The angle between the straight parts of the belt is 90°. Find the length of the belt.

9) A metric thread having a major diameter of 52 mm and a pitch of 5 mm is to be checked by the 3-wire method. Determine the best size of wire and, using this best wire size, determine the measurement over the wires if the thread is correct.

10) A metric thread having a major diameter of 30 mm and a pitch of 2 mm is to be checked using wires whose diameters are 1.14 mm. Calculate the measurement over the wires that will be obtained if the thread is correct.

# AREA OF A TRIANGLE

Three formulae are commonly used for finding the areas of triangles:

(1) If given the base and the altitude (i.e. vertical height).

(2) If any given two sides and the included angle.

(3) If given the three sides.

**Case (1)  Given the base and the altitude**

In Fig. 11.49,

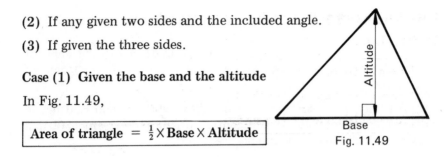

Area of triangle $= \frac{1}{2} \times$ **Base** $\times$ **Altitude**

Fig. 11.49

## EXAMPLE 11.16

Find the areas of the triangles shown in Fig. 11.50.

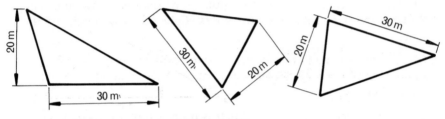

Fig. 11.50

In each case the 'base' is taken as the side of given length and the 'altitude' is measured perpendicular to this side.

Hence

$$\text{Triangular area} = \frac{1}{2} \times \text{Base} \times \text{Altitude}$$
$$= \frac{1}{2} \times 30 \times 20 = 300\,\text{m}^2 \quad \text{in each case}$$

## EXAMPLE 11.17

A trapezium is shown in Fig. 11.51 in which AB is parallel to CD. Find its area.

If we join AD then the trapezium is divided into two triangles, the 'bases' and 'altitudes' of which are known.

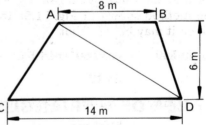

Fig. 11.51

$$\therefore \quad \text{Area of trapezium} = \text{Area of } \triangle ABD + \text{Area of } \triangle ADC$$
$$= \frac{1}{2} \times 8 \times 6 + \frac{1}{2} \times 14 \times 6 = 66\,\text{m}^2$$

## Case (2)  If given any two sides and the included angle

In Fig. 11.52,

| Area of triangle $= \frac{1}{2}bc \sin A$ |
|---|
| or    Area of triangle $= \frac{1}{2}ac \sin B$ |
| or    Area of triangle $= \frac{1}{2}ab \sin C$ |

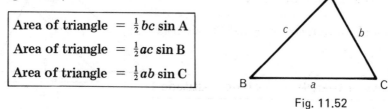

Fig. 11.52

## EXAMPLE 11.18

Find the area of the triangle shown in Fig. 11.53.

$$\text{Area} = \frac{1}{2} \times a \times c \times \sin B$$
$$= \frac{1}{2} \times 4 \times 3 \times \sin 30°$$
$$= 3 \text{ m}^2$$

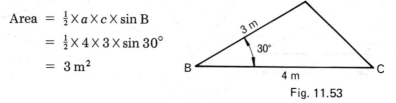

Fig. 11.53

## EXAMPLE 11.19

Find the area of the triangle shown in Fig. 11.54.

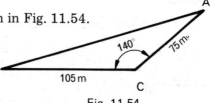

Fig. 11.54

$$\text{Area} = \frac{1}{2}ab \sin C$$
$$= \frac{1}{2} \times 105 \times 75 \times \sin 140°$$

We find the value of $\sin 140°$ by using
the method shown in Fig. 11.55 from
which it may be seen that

$$\sin 140° = \sin (180° - 140°)$$
$$= \sin 40°$$

$$\therefore \quad \text{Area} = \frac{1}{2} \times 105 \times 75 \times \sin 40°$$
$$= 2530 \text{ m}^2$$

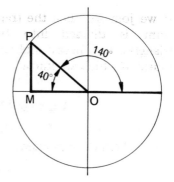

Fig. 11.55

Case (3) If given the three sides

In Fig. 11.56,

| | | |
|---|---|---|
| Area of triangle | = | $\sqrt{s(s-a)(s-b)(s-c)}$ |
| where $s$ | = | $\dfrac{a+b+c}{2}$ |

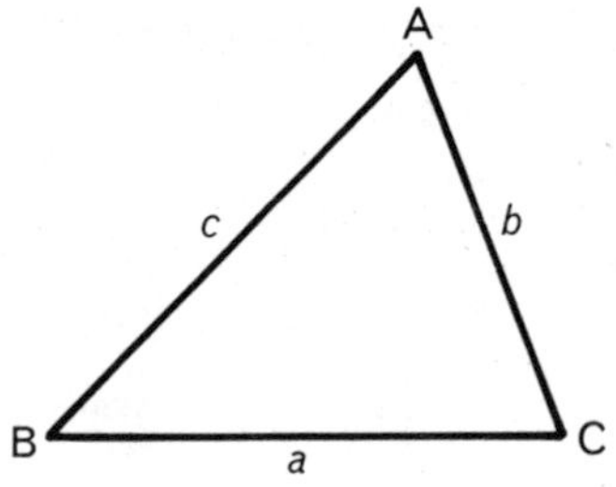


Fig. 11.56

**EXAMPLE 11.20**

A triangle has sides of lengths 3 m, 5 m and 6 m. What is its area?

Since we are given the lengths of 3 sides we use

$$\text{Area} = \sqrt{s(s-a)(s-b)(s-c)}$$

$$\text{Now, } s = \frac{3+5+6}{2} = 7$$

$$\therefore \text{ Area} = \sqrt{7\times(7-3)\times(7-5)\times(7-6)} = \sqrt{56} = 7.48\,\text{m}^2$$

**EXAMPLE 11.21**

The chain lines of a survey are shown in Fig. 11.57. Find the area of the plot of land.

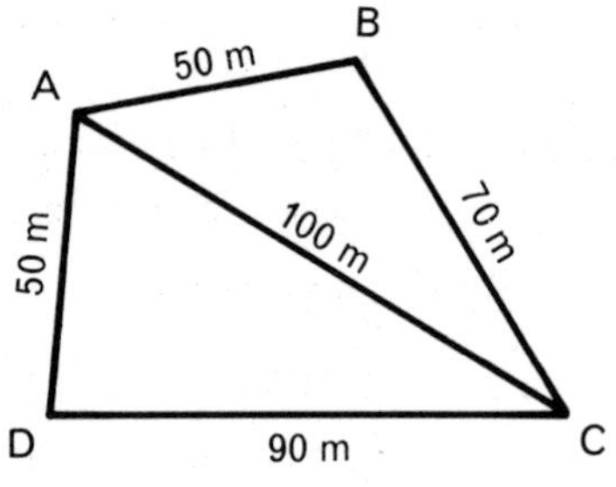


Fig. 11.57

The plot is the quadrilateral ABCD which is made up of the triangles ABC and ACD.

To find the area of $\Delta$ABC,

$$s = \frac{50+70+100}{2} = 110$$

$$\begin{aligned}\therefore \text{Area of } \Delta\text{ABC} &= \sqrt{s(s-a)(s-b)(s-c)} \\ &= \sqrt{110\times(110-50)\times(110-70)\times(110-100)} \\ &= \sqrt{110\times60\times40\times10} = 1625\,\text{m}^2\end{aligned}$$

To find the area of $\Delta$ACD,

$$s = \frac{50+90+100}{2} = 120$$

$\therefore$ Area of $\triangle ACD = \sqrt{s(s-a)(s-b)(s-c)}$

$\qquad\qquad = \sqrt{120\times(120-50)\times(120-90)\times(120-100)}$

$\qquad\qquad = \sqrt{120\times70\times30\times20} = 2245\,\text{m}^2$

$\therefore$ Area of quadrilateral $=$ Area of $\triangle ABC +$ Area of $\triangle ACD$

$\qquad\qquad\qquad\qquad = 1625 + 2245 = 3870\,\text{m}^2$

## Exercise 11.6

1) Find the area of a triangle whose base is 75 mm and whose altitude is 59 mm.

2) Find the area of an isosceles triangle whose equal sides are 82 mm and whose base is 95 mm.

3) A plate in the shape of an equilateral triangle has a mass of 12.25 kg. If the material has a mass of 3.7 kg/m², find the dimensions of the plate in mm.

4) Obtain the area of a triangle whose sides are 39.3 m and 41.5 m if the angle between them is 41°30′.

5) Find the area of the playground shown in Fig. 11.58.

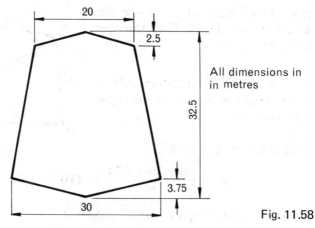

Fig. 11.58

6) Calculate the area of a triangle ABC if:
(a) $a = 4\,\text{m}$, $b = 5\,\text{m}$ and $\angle C = 49°$,
(b) $a = 3\,\text{m}$, $c = 6\,\text{m}$ and $\angle B = 63°44′$.

7) A triangle has sides 4 m, 7 m and 9 m long. What is its area?

8) A triangle has sides 37 mm, 52 mm and 63 mm long. What is its area?

9) Find the areas of the quadrilaterals shown in Fig. 15.59.

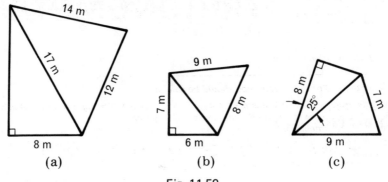

Fig. 11.59

10) Find the area of the triangle shown in Fig. 11.60.

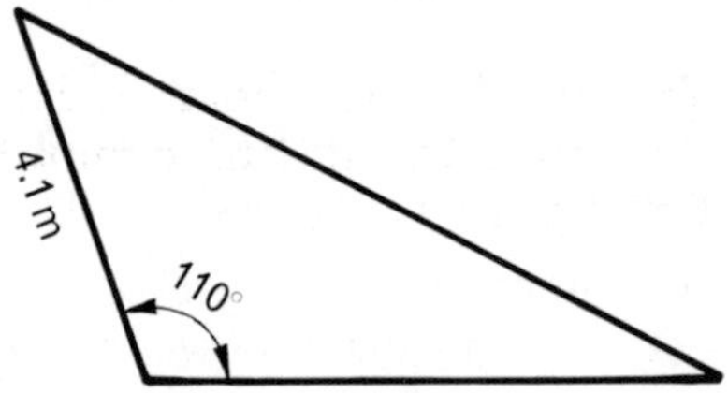

Fig. 11.60

11) What is the area of a parallelogram whose base is 7 m long and whose vertical height is 4 m?

12) Obtain the area of a parallelogram if two adjacent sides measure 112.5 mm and 105 mm and the angle between them is 49°.

13) Determine the length of the side of a square whose area is equal to that of a parallelogram with a 3 m base and a vertical height of 1.5 m.

14) Find the area of a trapezium whose parallel sides are 75 mm and 82 mm long respectively and whose vertical height is 39 mm.

15) Find the area of a regular hexagon,
(a) which is 4 m wide across flats,
(b) which has sides 5 m long.

16) Find the area of a regular octagon,
(a) which is 2 m wide across flats,
(b) which has sides 2 m long.

17) The parallel sides of a trapezium are 12 m and 16 m long. If its area is 220 m² what is its altitude?

 # SOLID TRIGONOMETRY

---

*After reaching the end of this chapter you should be able to:*

1.  Define the angle between a line and a plane.
2.  Define the angle between two relevant planes in a given three-dimensional problem.
3.  Solve three-dimensional triangulation problems

capable of being specified within a rectangular prism.
4.  Relate lengths and areas on an inclined plane to corresponding lengths and areas on plan.

---

## LOCATION OF A POINT IN SPACE

The location of a point in space requires co-ordinates referred to the origin O, and three mutually perpendicular axes $Ox$, $Oy$ and $Oz$ (Fig. 12.1).

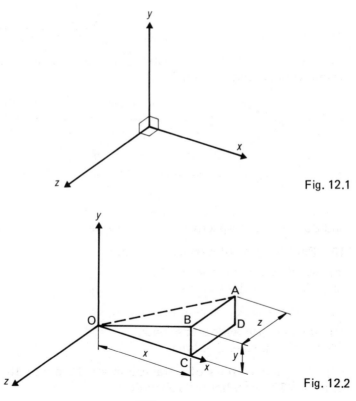

Fig. 12.1

Fig. 12.2

Consider the point A (Fig. 12.2). Let the co-ordinates measured parallel to the axes $Ox$, $Oy$ and $Oz$ be $x$, $y$ and $z$ respectively. To calculate the true lengths of the line OA we use the theorem of Pythagoras.

In the right-angled triangle OBC we have

$$OB^2 \;=\; OC^2 + BC^2 \;=\; x^2 + y^2$$

and in the right-angled triangle OAB we have

$$OA^2 \;=\; OB^2 + AB^2 \;=\; x^2 + y^2 + z^2$$

$$\therefore \qquad OA \;=\; \sqrt{x^2 + y^2 + z^2}$$

### EXAMPLE 12.1

Fig. 12.3 shows part of a hipped roof. Find the length of the hip rafter AB and the common rafter BD.

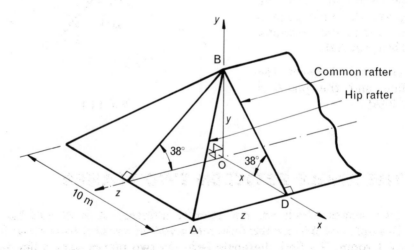

Fig. 12.3

In $\triangle BOD$,   $OD = 5$   and   $\angle BDO = 38°$

thus        $\dfrac{OB}{OD} = \tan \angle BDO$

$\therefore$        $OB = OD \tan \angle BDO = 5 \tan 38° = 3.906$

We now have the values $x = 5$ and $y = 3.906$

Thus    BD $= \sqrt{x^2 + y^2} = \sqrt{5^2 + 3.906^2} = \sqrt{40.257} = 6.345$

Hence the length of the common rafter is 6.345 m.

For the line AB we have  $x = 5$,  $y = 3.906$  and  $z = 5$

Thus

AB $= \sqrt{x^2 + y^2 + z^2} = \sqrt{5^2 + 3.906^2 + 5^2} = \sqrt{65.257} = 8.08\,\text{m}$

Hence the length of the hip rafter is 8.08 m.

# THE ANGLE BETWEEN A LINE AND A PLANE

In Fig. 12.4 the line AP intersects the $xz$ plane at A. To find the angle between AP and the plane, draw PM perpendicular to the plane and then join AM.

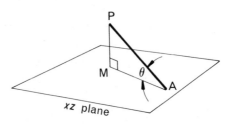

The angle between the line and the plane is $\angle$PAM.

Fig. 12.4

# THE ANGLE BETWEEN TWO PLANES

Two planes which are not parallel intersect in a straight line. Examples of this are the floor and a wall of a room, and two walls of a room. To find the angle between two planes draw a line in each plane which is perpendicular to the common line of intersection. The angle between the two lines is the same as the angle between the two planes.

Three planes usually intersect at a point as, for instance, two walls and the floor of a room.

Problems with solid figures are solved by choosing suitable right-angled triangles in different planes. It is essential to make a clear three-dimensional drawing in order to find these triangles. The examples which follow show the methods that should be adopted.

**EXAMPLE 12.2**

Fig. 12.5 shows a cuboid. Find the angle between the diagonal AG and the plane EFGH.

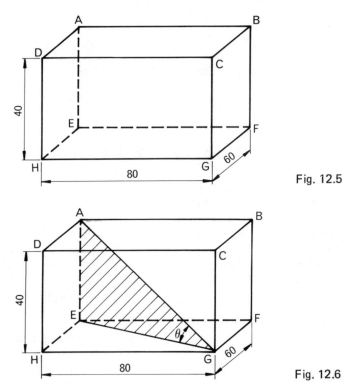

Fig. 12.5

Fig. 12.6

In order to find the required angle ($\theta$ in Fig. 12.6) we must use the right-angled triangle AEG, GE being the diagonal of the base rectangle.

In $\triangle$EFG, EF = 80 mm,  GF = 60 mm  and  EFG = 90°. Using Pythagoras' theorem gives

$$EG^2 = EF^2 + GF^2 = 80^2 + 60^2 = 10\,000$$

$$EG = \sqrt{10\,000} = 100\,\text{mm}$$

In $\triangle$AEG,  $\tan\theta = \dfrac{AE}{EG} = \dfrac{40}{100} = 0.4$

$$\theta = 21.8°$$

Hence the angle between the diagonal AG and the plane EFGH is 21.8°.

**EXAMPLE 12.3**

Fig. 12.7 shows a pyramid with a square base. The base has sides 60 mm long and the edges of the pyramid, VA, VB, VC and VD are each 100 mm long. Find the altitude of the pyramid.

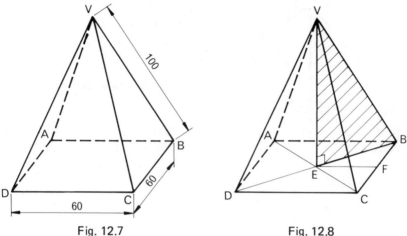

Fig. 12.7                    Fig. 12.8

The right-angled triangle VBE (Fig. 12.8) allows the altitude VE to be found, but first we must find BE from the right-angled triangle BEF.

In $\triangle BEF$, $BF = EF = 30$ mm and $\angle BFE = 90°$.

Using Pythagoras' theorem gives

$$BE^2 = BF^2 + EF^2 = 30^2 + 30^2 = 900 + 900 = 1800$$
$$BE = \sqrt{1800} = 42.43 \text{ mm}$$

In $\triangle VBE$, $BE = 42.43$ mm, $VB = 100$ mm and $\angle VEB = 90°$.

Using Pythagoras' theorem gives

$$VE^2 = VB^2 - BE^2 = 100^2 - 42.43^2$$
$$= 10\,000 - 1800 = 8200$$
$$VE = \sqrt{8200} = 90.6 \text{ mm}$$

# PROJECTED AREAS

In Fig. 12.9, the edges AD and BC of the rectangle ABCD are inclined at an angle $\theta$ to the horizontal plane. The rectangle $abcd$ represents the plane of ABCD.

$$\text{Area of rectangle ABCD} = AB \times BC$$

But    $\dfrac{bc}{BC} = \cos\theta$   or   $BC = \dfrac{bc}{\cos\theta}$,   and   $AB = ab$

∴    Area of rectangle ABCD $= ab \times \dfrac{bc}{\cos\theta}$

$$= \frac{ab \times bc}{\cos\theta}$$

$$= \frac{\text{Area of rectangle } abcd}{\cos\theta}$$

$$= \frac{\text{Projected plan area of ABCD}}{\cos\theta}$$

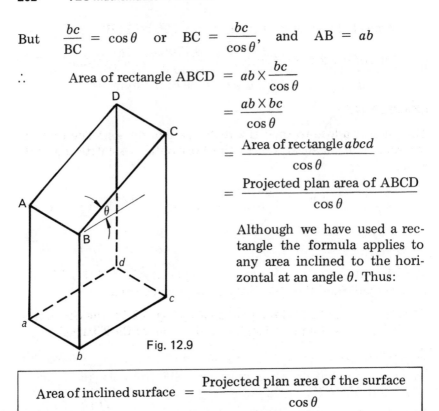

Although we have used a rectangle the formula applies to any area inclined to the horizontal at an angle $\theta$. Thus:

Fig. 12.9

---

Area of inclined surface $= \dfrac{\text{Projected plan area of the surface}}{\cos\theta}$

---

### EXAMPLE 12.4

A vertical chimney stack passes through a roof which is inclined at 41° to the horizontal (Fig. 12.10). The diameter of the stack is 150 mm. Find the area at the place where the stack passes through the roof.

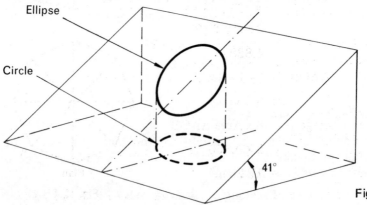

Fig. 12.10

Projected plan area of stack $= \pi \times 75^2 = 17700\,\text{mm}^2$

Area of stack at roof surface $= \dfrac{17700}{\cos 41^\circ} = 23400\,\text{mm}^2$

## EXAMPLE 12.5

The plan of a lean-to shed is a rectangle 3 m by 4 m. The roof of the shed is inclined at $22^\circ$ to the horizontal. Calculate the roof area.

$$\text{Roof area} = \frac{\text{Plan area}}{\cos \theta} = \frac{3 \times 4}{\cos 22^\circ} = 12.9\,\text{m}^2$$

## EXAMPLE 12.6

The plan of a steeple is a regular octagon of side 4 m. The steeple is 20 m high. Find the area of the roof to be covered.

The first step is to find the plan area. In Fig. 12.11, the octagon is divided into eight equal triangles. AOB is one of these triangles.

In $\triangle$ AOB,

$$\angle AOB = \frac{360^\circ}{8} = 45^\circ$$

Since $\triangle$ AOB is isosceles,

$$\angle AON = \frac{45^\circ}{2} = 22^\circ 30'$$

Now $\dfrac{ON}{AN} = \cot 22^\circ 30'$

$\therefore$ $\quad ON = AN \cot 22^\circ 30'$

$\qquad = 2 \cot 22^\circ 30'$

$\qquad = 4.828\,\text{m}$

Area $\quad \triangle AOB = \frac{1}{2} \times 4 \times 4.828$

$\qquad = 9.656\,\text{m}^2$

Area of octagon $= 8 \times$ Area $\triangle AOB$

$\qquad = 8 \times 9.656$

$\qquad = 77.25\,\text{m}^2$

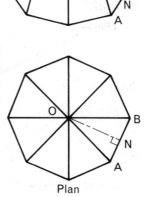

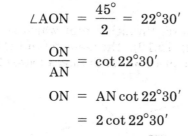

Plan

Fig. 12.11

To find the angle of slope of the roof the triangle PON is used (Fig. 12.11).

$$\tan \theta = \frac{OP}{ON} = \frac{20}{4.828} \qquad \therefore \quad \theta = 76.42°$$

Thus   Roof area $= \dfrac{\text{Plan area}}{\cos \theta} = \dfrac{77.25}{\cos 76.42°} = 329 \, \text{m}^2$

## HEIGHTS OF INACCESSIBLE OBJECTS

The height of a hill, tower, etc. cannot usually be measured directly. However if the angle of elevation to the top of the features is measured from two points whose distance apart is known, the height of the feature can be determined. When the two points are in line with the feature we proceed as shown on page 169. We now deal with the case of the two points not being in line.

### EXAMPLE 12.7

The angle of elevation of a small hill from a point due south of the hill is 23°. The angle of elevation to the top of the same hill from a points are in line with the factors we proceed as shown on page 169. We now deal with the case of the two points not being in line.

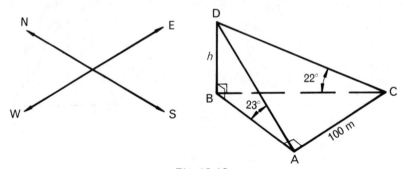

Fig. 12.12

In Fig. 12.12, BD represents the hill. If we let  BD $= h$,  then:

In $\triangle$ABD, $\angle$DBA $= 90°$ and $\angle$DAB $= 23°$

$\therefore \qquad\qquad$ AB $= h \cot 23° = 2.356h$

In $\triangle$BCD, $\angle$DBC $= 90°$ and $\angle$BCD $= 22°$

$\therefore \qquad\qquad$ BC $= h \cot 22° = 2.475h$

In $\triangle ABC$, $\angle BAC = 90°$.

Using Pythagoras' theorem   $BC^2 = AB^2 + AC^2$

then                       $(2.475h)^2 = (2.356h)^2 + 100^2$

or                          $6.126h^2 = 5.551h^2 + 10\,000$

from which                     $h = 132$

Hence the height of the hill is $132$ metres.

# Exercise 12.1

**1)** Fig. 12.13 shows a cuboid.

(a)  Sketch the rectangle EFGH.
(b)  Calculate the diagonal FH of rectangle EFGH.
(c)  Sketch the rectangle FHDB adding known dimensions.
(d)  Calculate the diagonal BH of rectangle FHDB.

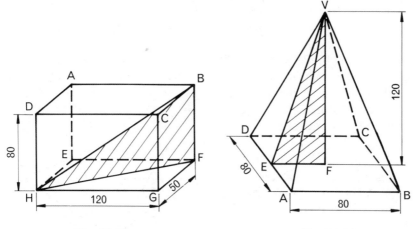

Fig. 12.13                    Fig. 12.14

**2)** Fig. 12.14 shows a pyramid on a square base of side 80 mm. The altitude of the pyramid is 120 mm.

(a)  Calculate EF.
(b)  Draw the triangle VEF adding known dimensions.
(c)  Find the angle VEF.
(d)  Calculate the slant height VE.
(e)  Calculate the area of $\triangle VAD$.
(f)  Calculate the complete surface area of the pyramid.

3) Fig. 12.15 shows a pyramid on a rectangular base. Calculate the length VA.

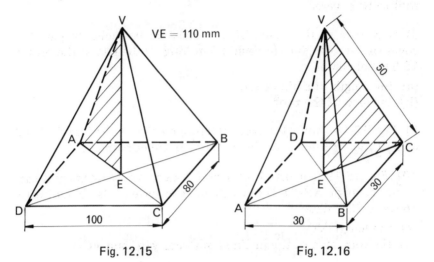

Fig. 12.15                    Fig. 12.16

4) Fig. 12.16 shows a pyramid on a square base with

$$VA = VB = VC = VD = 50\,mm$$

Calculate the altitude of the pyramid.

5) A hipped roof makes an angle of $42°$ with the horizontal and its span is $12\,m$. Calculate the length of a common rafter and the length of a hip rafter.

6) In Fig. 12.17, ABCD represents part of a hill-side. A line of greatest slope AB is inclined at $36°$ to the horizontal AE and runs due north of A. The line AF bears $050°$ (N50°E) and C is $2500\,m$ east of B. The lines of BE and CF are vertical. Calculate:

(a) the height of C above A,

(b) the angle between AB and AC.

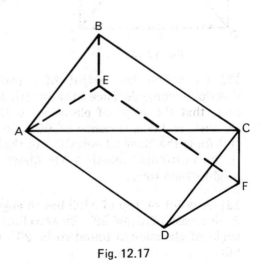

Fig. 12.17

**7)** A lean-to shed has a base which is 6 m long by 3 m wide. The roof makes an angle of 24° to the horizontal. Calculate the area of roof to be covered.

**8)** A roof is in the shape of an octagonal pyramid. In plan the edges of the octagon are each 4.2 m long. If the rise of the roof is 12.8 m calculate:

(a) the length of a hip rafter,
(b) the area of the roof.

**9)** Fig. 12.18 shows the plan of a hipped roof. Calculate the total area of the roof if it is inclined at 40° to the horizontal.

**10)** Fig. 12.19 is a sketch of a roof feature whose horizontal base is the rectangle ABCD 37 m by 23 m. The vertex V is 12 m directly above A. Calculate:

(a) the length VC,
(b) the total area of the inclined surfaces VBC and VCD.

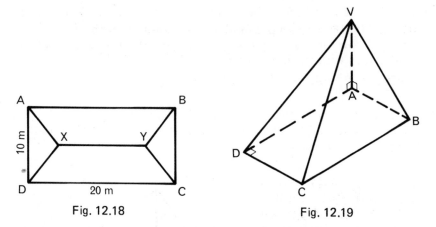

Fig. 12.18          Fig. 12.19

**11)** To obtain the height of a tower a surveyor sets up his theodolite some distance to the north of the base of the tower and finds that the angle of elevation to the top of the tower is 31°. He then moves a distance of 60 m to the west of the first point and finds the angle of elevation to the top of the tower to be 22°. If the instrument stands 1.5 m above ground level calculate the height of the tower.

**12)** A point on top of a hill has an angle of elevation from a point P, due east of it, of 29°. By travelling 180 m due north of P the angle of elevation is found to be 26°. Calculate the height of the hill.

**13)** Fig. 12.20 shows the plan and elevation of a north light roof. Calculate the area of the roof surfaces.

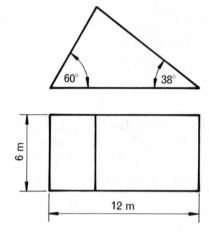

Fig. 12.20

**14)** A spire is covered with sheet copper. Its plan is a regular hexagon of side 2 m. Each sloping face is inclined at $76°$ to the horizontal. Calculate the area of the copper covering.

**15)** Part of a hillside is inclined uniformly at an angle of $39°$ to the horizontal. A straight road on the hillside makes an angle of $58°$ with the line of greatest slope. Determine the angle that the road makes with the horizontal.

# 13.

# CONTINUED FRACTIONS

*fter reaching the end of this chapter you should be able to:*

*1. Describe the form of a continued fraction.*
*2. Define the terms quotients and convergents.*
*3. Calculate the convergents of a given fraction by successive division and by using a tabulation method.*

*4. Use the final convergent as a check.*
*5. Calculate the difference between a fraction and a convergent.*
*6. Apply continued fractions to practical situations.*

## USE AND METHOD

The method of continued fractions is used principally for arranging gears in compound trains, such as may be needed when cutting threads on a lathe. Suppose that for cutting a thread the following ratio is required:

$$\frac{\text{Driver}}{\text{Driven}} = \frac{31}{71}$$

Now both 31 and 71 are prime numbers and hence special wheels having 31 teeth and 71 teeth are required for cutting the thread. In many cases, such as this, either the numerator, or the denominator, or both will not factorise. It is therefore necessary to obtain other fractions whose values are closed approximations to the given ratio and whose numerators and denominators will factorise. The method of continued fractions will give a ratio which is often close enough to that required to allow the standard change wheels to be used.

Any vulgar fraction whose value is less than 1 can be put in the form

$$\cfrac{1}{a_1 + \cfrac{1}{a_2 + \cfrac{1}{a_3}}} \quad \text{etc.,}$$

by choosing appropriate values for $a_1, a_2, a_3$, etc. A fraction written in this form is called a continued fraction. The values of $a_1, a_2, a_3$, etc., are obtained by a special method of division, which is described later. Hence they are called partial quotients, $a_1$, being the first, $a_2$ the second, $a_3$ the third, etc.

For the purpose of explaining the method we will convert 31/71 to a continued fraction. The way in which the partial quotients are obtained is as follows:

**(1)** To obtain the first partial quotient divide the numerator into the denominator:

$$31)\overline{71}(2 \ = \ \text{First partial quotient}$$
$$\underline{62}$$
$$9 \ = \ \text{Remainder.}$$

**(2)** To obtain the second partial quotient divide the remainder into the previous divisor:

$$9)\overline{31}(3 \ = \ \text{Second partial quotient}$$
$$\underline{27}$$
$$4 \ = \ \text{Remainder.}$$

**(3)** The remaining partial quotients are obtained in a similar way, the remainder being divided, in each case, into the previous divisor. The calculation of the partial quotients ends when the remainder becomes zero.

The entire calculation is best done in the following way:

$$31)\overline{71}(2 \qquad \ldots \text{Divide the numerator into the denominator.}$$
$$\underline{62}$$
$$9)\overline{31}(3 \qquad \ldots \text{Divide the remainder into the original}$$
$$\underline{27} \qquad \qquad \text{divisor.}$$
$$4)\overline{9}(2 \qquad \ldots \text{Divide the remainder into the previous}$$
$$\underline{8} \qquad \qquad \text{divisor, and so on until no remainder is left.}$$
$$1)\overline{4}(4$$
$$\underline{4}$$
$$0$$

The partial quotients are, therefore, $a_1 = 2$, $a_2 = 3$, $a_3 = 2$ and $a_4 = 4$. Thus,

$$\frac{31}{71} = \cfrac{1}{2 + \cfrac{1}{3 + \cfrac{1}{2 + \cfrac{1}{4}}}}$$

which may be verified by ordinary arithmetic.

In order to obtain approximations to the original fraction, the continued fraction may be expressed as a series by stating the values of

$$\frac{1}{a_1}, \quad \frac{1}{a_1+\dfrac{1}{a_2}}, \quad \frac{1}{a_1+\dfrac{1}{a_2+\dfrac{1}{a_3}}} \quad , \quad \text{etc.}$$

In the case of the fraction 31/71 the various approximations are

$$\frac{31}{71} \simeq \frac{1}{2}$$

$$\frac{31}{71} \simeq \frac{1}{2+\dfrac{1}{3}} = \frac{1}{7/3} = \frac{3}{7}$$

$$\frac{31}{71} \simeq \frac{1}{2+\dfrac{1}{3+\dfrac{1}{2}}} = \frac{1}{2+\dfrac{1}{7/2}} = \frac{1}{2+\dfrac{2}{7}} = \frac{1}{16/7} = \frac{7}{16}$$

In order to compare the values of the above approximations the table below has been compiled. The correct value, to four significant figures, of 31/71 is 0.4366. If the value of the approximation is greater than this then the error is said to be positive, if less it is said to be negative.

| Partial quotient number | Fraction | Decimal equivalent | Error | % Error |
|---|---|---|---|---|
| 1 | 1/2 | 0.5000 | + 0.0634 | 14.52 |
| 2 | 3/7 | 0.4286 | − 0.0080 | − 1.83 |
| 3 | 7/16 | 0.4375 | + 0.0009 | 0.21 |
| 4 | 31/71 | 0.4366 | 0 | — |

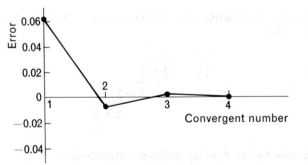

Fig. 13.1

It will be noticed that successive approximations are alternately too large and too small. The higher the number of the partial quotient, the nearer it approaches the correct value of the original fraction. The approximations therefore converge on the original value of the fraction and the approximations are known as convergents. Thus for the fraction 31/71, 1/2 is the first convergent, 3/7 is the second, 7/16 is the third and 31/71 is the fourth convergent. A graph of error against convergent number is shown in Fig. 13.1 and it illustrates the manner in which the convergents converge on the original fraction.

## QUICK METHOD FOR OBTAINING CONVERGENTS

The method of obtaining the successive convergents from the partial quotients which has been shown previously is cumbersome and can easily lead to errors. A far better method is the tabular one, which will now be described by considering the fraction 31/71. It will be remembered that the partial quotients were 2, 3, 2 and 4. The following table is now drawn up:

| Convergent number | | 1 | 2 | 3 | 4 |
|---|---|---|---|---|---|
| Partial quotient | | 2 | 3 | 2 | 4 |
| 1 | 0 | | | | |
| 0 | 1 | | | | |

To start off the process we need the two initial convergents 1/0 and 0/1, which are written into the table as shown. (Irrespective of the values of the partial quotients the values 1/0 and 0/1 are always used to initiate the process.)

The rule for finding the numerator of any convergent is:

*Multiply the partial quotient appropriate to the convergent by the numerator of the preceding convergent, and add the numerator of the previous convergent.*

Thus, the numerator of the first convergent is $2 \times 0 + 1 = 1$. The denominators are found in a similar way, but we now see the denominators of the preceding and second preceding denominators.

The denominator of the first convergent is therefore $2 \times 1 + 0 = 2$. The following table shows all the calculations necessary in finding the convergents.

| Convergent number | | | 1 | 2 | 3 | 4 |
|---|---|---|---|---|---|---|
| Partial quotient | | | 2 | 3 | 2 | 4 |
| | 1 | 0 | $2 \times 0 + 1$ $= 1$ | $3 \times 1 + 0$ $= 3$ | $2 \times 3 + 1$ $= 7$ | $4 \times 7 + 3$ $= 31$ |
| | 0 | 1 | $2 \times 1 + 0$ $= 2$ | $3 \times 2 + 1$ $= 7$ | $2 \times 7 + 2$ $= 16$ | $4 \times 16 + 7$ $= 71$ |

From the table it can be seen that the convergents are, as before, 1/2, 3/7, 7/16 and 31/71.

The method is self-checking because if all the calculation has been performed correctly the last convergent is the same as the original fraction or the original fraction reduced to its lowest terms. Thus, if the fraction 195/327 were expressed as a continued fraction the last convergent would be 65/109.

## EXAMPLE 13.1

Find the successive
convergents of 126/233

```
126)233(1
    126
    107)126(1
        107
        19)107(5
           95
           12)19(1
              12
              7)12(1
                7
                5)7(1
                  5
                  2)5(2
                    4
                    1)2(2
                      2
                      0
```

|   |   | 1 | 1 | 5 | 1 |
|---|---|---|---|---|---|
| 1 | 0 | $1 \times 0 + 1$ $= 1$ | $1 \times 1 + 0$ $= 1$ | $5 \times 1 + 1$ $= 6$ | $1 \times 6 + 1$ $= 7$ |
| 0 | 1 | $1 \times 1 + 0$ $= 1$ | $1 \times 1 + 1$ $= 2$ | $5 \times 2 + 1$ $= 11$ | $1 \times 11 + 2$ $= 13$ |

|   | 1 | 1 | 2 | 2 |
|---|---|---|---|---|
| Table cont. | $1 \times 7 + 6$ $= 13$ | $1 \times 13 + 7$ $= 20$ | $2 \times 20 + 13$ $= 53$ | $2 \times 53 + 20$ $= 126$ |
|   | $1 \times 13 + 11$ $= 24$ | $1 \times 24 + 13$ $= 37$ | $2 \times 37 + 24$ $= 98$ | $2 \times 98 + 37$ $= 233$ |

The successive convergents are, therefore,

$$\frac{1}{1}, \frac{1}{2}, \frac{6}{11}, \frac{7}{13}, \frac{13}{24}, \frac{20}{37}, \frac{53}{98} \text{ and } \frac{126}{233}$$

# IMPROPER FRACTIONS

### EXAMPLE 13.2

Find the successive convergents of 112/31

Now 112/31 may be regarded as $\dfrac{1}{31/112}$

Thus if we find the successive convergents of 31/112 we can easily find the successive convergents of 112/31.

As before we must first find the partial quotients.

```
31)112(3
   93
   19)31(1
      19
      12)19(1
         12
          7)12(1
             7
             5)7(1
                5
                2)5(2
                   4
                   1)2(2
                      2
                      0
```

Thus the partial quotients are 3, 1, 1, 1, 1, 2 and 2. The table for finding the convergents of 31/112 is now drawn up:

|   |   | 3 | 1 | 1 | 1 | 1 | 2 | 2 |
|---|---|---|---|---|---|---|---|---|
| 1 | 0 | 1 | 1 | 2 | 3 | 5 | 13 | 31 |
| 0 | 1 | 3 | 4 | 7 | 11 | 18 | 47 | 112 |

The successive convergents of 31/112 are therefore

$$\frac{1}{3}, \frac{1}{4}, \frac{2}{7}, \frac{3}{11}, \frac{5}{18}, \frac{13}{47} \text{ and } \frac{31}{112}$$

The convergents of 112/31 are the reciprocals of these, so by inverting each we get

$$\frac{3}{1}, \frac{4}{1}, \frac{7}{2}, \frac{11}{3}, \frac{18}{5}, \frac{47}{13} \text{ and } \frac{112}{31}$$

# DECIMAL NUMBERS

When we require the continued fraction of a decimal number, we must first change the decimal to a fraction.

**EXAMPLE 13.3**

Find the successive convergents of 0.358

Now $\quad 0.358 = \dfrac{358}{1000} = \dfrac{179}{500}$

Note that before attempting to find the successive quotients the initial fraction 358/1000 should be reduced to its lowest terms.

$$179)500(2$$
$$\underline{358}$$
$$142)179(1$$
$$\underline{142}$$
$$37)142(3$$
$$\underline{111}$$
$$31)37(1$$
$$\underline{31}$$
$$6)31(5$$
$$\underline{30}$$
$$1)6(6$$
$$\underline{6}$$
$$0$$

Hence the successive quotients are 2, 1, 3, 1, 5 and 6

To find the successive convergents a table is drawn up in the usual way:

| | | 2 | 1 | 3 | 1 | 5 | 6 |
|---|---|---|---|---|---|---|---|
| 1 | 0 | 1 | 1 | 4 | 5 | 29 | 179 |
| 0 | 1 | 2 | 3 | 11 | 14 | 81 | 500 |

The successive convergents are therefore

$$\frac{1}{2}, \frac{1}{3}, \frac{4}{11}, \frac{5}{14}, \frac{29}{81} \text{ and } \frac{179}{500}$$

# APPLICATIONS TO THE DIVIDING HEAD

At this stage the student will be familiar with the dividing head and its manner of use on the milling machine.

Now      40 turns $=$ 1 turn

$\therefore$          1 turn $= \frac{1}{40}$ turn $= 9°$ turn of the work

The Brown and Sharpe dividing head is provided with the indexing plates having the following hole circles:

$$15 \quad 18 \quad 20 \quad 23 \quad 27 \quad 31 \quad 37 \quad 41 \quad 47$$
$$16 \quad 17 \quad 19 \quad 21 \quad 29 \quad 33 \quad 39 \quad 43 \quad 49$$

## EXAMPLE 13.4

Using the Brown and Sharpe indexing plates shown above calculate the nearest indexing and the actual angle obtained for an angle of $32°17'$.

Since

1 turn of crank $= 9°$ turn of work

$$\text{Indexing} = \frac{32°17'}{9°} = \frac{32\frac{17}{60}}{9} = \frac{1937/60}{9} = \frac{1937}{540} = 3\frac{317}{540}$$

We now find the successive convergents of 317/540, which are

$$\frac{1}{1}, \frac{1}{2}, \frac{3}{5}, \frac{7}{12}, \frac{10}{17}, \frac{27}{46}, \frac{145}{247} \text{ and } \frac{317}{540}$$

Trying each of the convergents in turn, starting with the *highest* we see that 5th convergent, 10/17, is suitable since we have an indexing plate with a 17 hole circle. Hence suitable indexing is

3 full turns + 10 holes in a 17 circle

To find the actual angle obtained, call it $\theta°$.

Then
$$\frac{\theta°}{9°} = 3\tfrac{10}{17} = \frac{61}{17}$$

∴
$$\theta = \frac{61 \times 9}{17} = \frac{549}{17} = 32\tfrac{5}{17}° = 32°17'39''$$

## APPLICATION TO SPIRAL MILLING

### EXAMPLE 13.5

In spiral milling with the gears as arranged as shown in Fig. 13.2 the ratio applies:

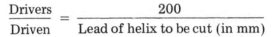

$$\frac{\text{Drivers}}{\text{Driven}} = \frac{200}{\text{Lead of helix to be cut (in mm)}}$$

With the Brown and Sharpe dividing head the following gears are supplied: two of 24 teeth, and one each of 28, 32, 40, 44, 48, 56, 64, 72, 86, and 100 teeth. Find suitable gearing to cut a helix having a lead of 326 mm and find the error introduced by using this gearing.

$$\frac{\text{Drivers}}{\text{Driven}} = \frac{200}{326} = \frac{100}{163}$$

This is an awkward fraction, and using continued fractions the convergents are as follows:

$$\frac{1}{1}, \frac{1}{2}, \frac{2}{3}, \frac{3}{5}, \frac{8}{13}, \frac{19}{31}, \frac{27}{44} \text{ and } \frac{100}{163}$$

The higher the convergent the greater will be the accuracy, so we will try using the seventh convergent for suitable gearing. Thus we have

$$\frac{\text{Drivers}}{\text{Driven}} = \frac{27}{44} = \frac{9 \times 3}{44 \times 1} = \frac{72 \times 24}{44 \times 64}$$

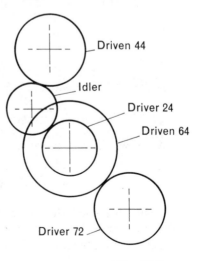

Driven 44

Idler

Driver 24

Driven 64

Driver 72

Fig. 13.2

Finally to find the error introduced, let $L$ be the lead of the helix that will be cut by the train calculated.

Then
$$\frac{27}{44} = \frac{200}{L}$$

$$L = \frac{44 \times 200}{27} = \frac{8800}{27} = 325.93 \text{ mm}$$

which is sufficiently close to 326 mm for our purposes, the percentage error being 0.02%.

## APPLICATION TO THREAD CUTTING

It is shown in books of Workshop Technology that in lathe work:

$$\frac{\text{Number of teeth in driving wheels}}{\text{Number of teeth in driven wheels}} = \frac{\text{Lead of thread to be cut}}{\text{Lead of leadscrew}}$$

Often the pitch of the thread to be cut is such that the standard change wheels cannot be used, and we then use the method of continued fractions. It should be remembered that the standard change wheels range from 20 teeth to 120 teeth in steps of 5 teeth.

### EXAMPLE 13.6

Using the standard change wheels find the nearest gear train to cut a thread with a lead of 2.62 mm. The lead of the lathe leadscrew is 5 mm.

$$\frac{\text{Drivers}}{\text{Driven}} = \frac{\text{Lead of screw to be cut}}{\text{Lead of leadscrew}} = \frac{2.62}{5} = \frac{262}{500} = \frac{131}{250}$$

Since 131 is a prime number, we obtain the successive convergents of 131/250 in the usual way. They are found to be:

$$\frac{1}{1}, \frac{1}{2}, \frac{10}{19}, \frac{11}{21} \text{ and } \frac{131}{250}$$

We see that the fourth convergent 11/21 allows the standard change wheels to be used since:

$$\frac{\text{Drivers}}{\text{Driven}} = \frac{11}{21} = \frac{11 \times 5}{21 \times 5} = \frac{55}{105}$$

To find the error using this approximation, let the lead be $L$ as obtained using it.

Then
$$\frac{\text{Drivers}}{\text{Driven}} = \frac{L}{5}$$

$\therefore$
$$L = \frac{11 \times 5}{21} = 2.61905 \text{ mm}$$

Then Percentage error $= \dfrac{2.62 - 2.61905}{2.62} \times 100 = 0.036\%$

This is insignificant when the length of the thread is short, e.g. in a length of 262 mm, that is, 100 threads, the error is only 0.095 mm. However if the screw thread is very long this lead error can become very important. In cases of this kind we have to find intermediate values between the convergents. This is beyond the scope of our work at this stage.

## EXAMPLE 13.7

Using standard change wheels find a train which is satisfactory for cutting a thread having a lead of 1.86 mm. The lathe lead-screw has a lead of 5 mm.

We have

$$\frac{\text{Drivers}}{\text{Driven}} = \frac{\text{Lead of screw to be cut}}{\text{Lead of leadscrew}} = \frac{1.86}{5} = \frac{186}{500} = \frac{93}{250}$$

Now 93 will factorise to give $31 \times 3$, but 31 is still unsatisfactory because the smallest wheel we can use has $31 \times 5$ or 155 teeth, which is outside the range of the usual change wheels. We therefore obtain the successive convergents of 93/250 in the usual way. These are found to be

$$\frac{1}{2}, \frac{1}{3}, \frac{3}{8}, \frac{13}{35}, \frac{16}{43} \text{ and } \frac{93}{250}$$

The fifth convergent is no good because 43 is a prime number, so we must use the fourth, 13/35. Hence,

$$\frac{\text{Drivers}}{\text{Driven}} = \frac{13}{35} = \frac{6.5 \times 2}{7 \times 5} = \frac{(6.5 \times 10) \times (2 \times 20)}{(7 \times 10) \times (5 \times 20)} = \frac{65 \times 40}{70 \times 100}$$

This is, of course, a compound train.

## Exercise 13.1

1) Express the following fractions as continued fractions and find their convergents:

(a) $\dfrac{87}{127}$     (b) $\dfrac{37}{61}$     (c) $\dfrac{1343}{2792}$     (d) $\dfrac{373}{685}$

2) Find the successive convergents of the following improper fractions:

(a) $\dfrac{923}{119}$     (b) $\dfrac{89}{41}$     (c) $\dfrac{4382}{1743}$

3) Find the successive convergents of the following:

(a) 0.283     (b) 0.981     (c) 1.764     (d) 3.829

4) Show by the method of continued fractions that 22/7 is a suitable approximation to $\pi$ (take $\pi$ as 3.142).

5) Using the Brown and Sharpe indexing plates (page 216) calculate the nearest indexing possible and find the actual angle obtained for the following angles:

(a) 19.21°     (b) 48°33′     (c) 79°27′     (d) 105°11′

6) In cutting a helix on a milling machine the ratio

$$\frac{\text{Drivers}}{\text{Driven}} = \frac{200}{\text{Lead of helix to be cut (in mm)}}$$

is used. If the machine is equipped with gear wheels having 24, 28, 32, 40, 44, 48, 56, 64, 72, 86 and 100 teeth find suitable gearing for cutting a helix having a lead of 225.4 mm. What is the actual lead obtained?

7) Find suitable gearing for cutting a helix of 306.2 mm on a milling machine where the following ratio applies:

$$\frac{\text{Drivers}}{\text{Driven}} = \frac{200}{\text{Lead of helix to be cut (in mm)}}$$

A double set of the following gear wheels is available: 24, 28, 32, 40, 44, 48, 56, 64, 72, 86 and 100 teeth.

8) A thread having a lead of 5.38 mm is to be cut on a lathe having a 5 mm leadscrew. Find suitable change wheels for cutting the thread. What error in the lead is introduced by using these wheels?

**9)** Using the standard change wheels find a train which is satisfactory for cutting a thread having a lead of 2.06 mm. The lathe has a 5 mm leadscrew. Find the actual lead which, according to the calculations, will be cut.

**10)** A 3 BA thread (0.729 mm pitch) is to be cut on a lathe with a 5 mm leadscrew. Calculate the nearest possible ratio which can be obtained with the usual change wheels and find the error introduced by using these wheels.

 # QUADRATIC EQUATIONS

## THE PRODUCT OF TWO BINOMIAL EXPRESSIONS

A binomial expression consists of two terms. Thus $3x + 5$, $a + b$, $2x + 3y$ and $4p - q$ are all binomial expressions.

The product $(a + b)(c + d) = a(c + d) + b(c + d) = ac + ad + bc + bd$

It will be noticed that the expression on the right-hand side is obtained by multiplying each term in the one bracket by each term in the other bracket. The process is illustrated below:

$$(a + b)(c + d) = ac + ad + bc + bd$$

**EXAMPLE 14.1**

a) $(3x + 2)(4x + 5) = 3x \times 4x + 3x \times 5 + 2 \times 4x + 2 \times 5$

$$= 12x^2 + 15x + 8x + 10 = 12x^2 + 23x + 10$$

b) $(2p - 3)(4p + 7) = 2p \times 4p + 2p \times 7 - 3 \times 4p - 3 \times 7$

$$= 8p^2 + 14p - 12p - 21 = 8p^2 + 2p - 21$$

c) $(z - 5)(3z - 2) = z \times 3z + z \times (-2) - 5 \times 3z - 5 \times (-2)$

$$= 3z^2 - 2z - 15z + 10 = 3z^2 - 17z + 10$$

# THE SQUARE OF A BINOMIAL EXPRESSION

$$(a + b)^2 = (a + b)(a + b) = a^2 + ab + ba + b^2 = a^2 + 2ab + b^2$$

The square of a binomial expression is the sum of the squares of the two terms and twice their product.

$$(a - b)^2 = (a - b)(a - b) = a^2 - ab - ba + b^2 = a^2 - 2ab + b^2$$

### EXAMPLE 14.2

a) $(2x + 5)^2 = (2x)^2 + 2 \times 2x \times 5 + 5^2 = 4x^2 + 20x + 25$

b) $(3x - 2)^2 = (3x)^2 + 2 \times 3x \times (-2) + (-2)^2 = 9x^2 - 12x + 4$

# THE PRODUCT OF THE SUM AND DIFFERENCE OF TWO TERMS

$$(a + b)(a - b) = a^2 - ab + ba - b^2 = a^2 - b^2$$

This result is the difference of the squares of the two terms.

### EXAMPLE 14.3

a) $(8x + 3)(8x - 3) = (8x)^2 - 3^2 = 64x^2 - 9$

b) $(2x + 5y)(2x - 5y) = (2x)^2 - (5y)^2 = 4x^2 - 25y^2$

# Exercise 14.1

Find the products of the following:

1) $(x + 1)(x + 2)$         2) $(2x + 5)(x + 3)$

3) $(2x + 4)(3x + 2)$       4) $(x - 4)(x - 2)$

5) $(x - 2)(3x - 5)$        6) $(3x - 1)(2x - 5)$

7) $(x + 3)(x - 1)$         8) $(x - 5)(x + 3)$

9) $(3x - 5)(x + 6)$        10) $(6x - 7)(2x + 3)$

11) $(2p - q)(p - 3q)$      12) $(3v + 2u)(2v - 3u)$

13) $(x + 3)(x - 3)$        14) $(2x + 3)(2x - 3)$

15) $(x + 1)^2$             16) $(2x + 3)^2$

17) $(x - 1)^2$             18) $(2x - 3)^2$

19) $(2a + 3b)^2$           20) $(x + y)^2$

21) $(a - b)^2$             22) $(3x - 4y)^2$

# QUADRATIC EXPRESSIONS

A quadratic expression is one in which the highest power of the symbol used is the square. Typical examples are $x^2 - 5x + 3$ or $3x^2 - 9$ in which there is no power of $x$ greater than $x^2$.

You will see, from the work in the previous section, that when two binomial expressions are multiplied together the result is always a quadratic expression.

It is often necessary to try and reverse this procedure. This means that we start with a quadratic expression and wish to express this as the product of two binomial expressions—this is not always possible. For example the expressions $x^2 + 1$ or $a^2 + b^2$ cannot be factorised. You may check this for yourself after following the next section of this chapter.

# FACTORISING QUADRATIC EXPRESSIONS

Consider $(7x + 4)(2x + 3)$

Now $(7x + 4)(2x + 3) = 14x^2 + 21x + 8x + 12$

$$= 14x^2 + 29x + 12$$

The following points should be noted:

**(1)** The first terms in each bracket when multiplied together give the first term of the quadratic expression.

**(2)** The middle term of the quadratic expression is formed by multiplying together the terms connected by a line (see above equation) and then adding them together.

**(3)** The last terms in each bracket when multiplied together give the last term of the quadratic expression.

In most cases, when factorising a quadratic expression, we find all the possible factors of the first and last terms. Then, by trying various combinations, the combination which gives the correct middle term may be found.

**EXAMPLE 14.4**

Factorise $2x^2 + 5x - 3$

| Factors of $2x^2$ | | Factors of $-3$ | |
|---|---|---|---|
| $2x$ | $x$ | $-3$ | $+1$ |
| | | $+3$ | $-1$ |

Combinations of these factors are:

$$(2x-3)(x+1) = 2x^2-x-3 \quad \text{which is incorrect,}$$
$$(2x+1)(x-3) = 2x^2-5x-3 \quad \text{which is incorrect,}$$
$$(2x+3)(x-1) = 2x^2+x-3 \quad \text{which is incorrect,}$$
$$(2x-1)(x+3) = 2x^2+5x-3 \quad \text{which is correct.}$$

Hence $\qquad 2x^2+5x-3 = (2x-1)(x+3)$

**EXAMPLE 14.5**

Factorise $12x^2-35x+8$

| Factors of $12x^2$ | | Factors of 8 | |
|:---:|:---:|:---:|:---:|
| $12x$ | $x$ | 1 | 8 |
| $6x$ | $2x$ | 8 | 1 |
| $3x$ | $4x$ | $-1$ | $-8$ |
| | | $-8$ | $-1$ |
| | | 2 | 4 |
| | | 4 | 2 |
| | | $-2$ | $-4$ |
| | | $-4$ | $-2$ |

By trying each combination in turn the only one which will produce the correct middle term of $-35x$ is found to be $(3x-8)(4x-1)$.

$$\therefore \qquad 12x^2-35x+8 = (3x-8)(4x-1)$$

# Where the Factors form a Perfect Square

A quadratic expression, which factorises into the product of two identical brackets resulting in a perfect square, may be factorised by the method used previously. However, if you can recognise that the result will be a perfect square then the problem becomes easier.

It has been shown that

$$(a+b)^2 = a^2+2ab+b^2 \quad \text{and} \quad (a-b)^2 = a^2-2ab+b^2$$

The square of a binomial expression therefore consists of:

(Square of 1st term) + (Twice product of terms)
                    + (Square of 2nd term)

**EXAMPLE 14.6**

Factorise $9a^2 + 12ab + 4b^2$

Now  $9a^2 = (3a)^2$,  $4b^2 = (2b)^2$  and  $12ab = 2 \times 3a \times 2b$

$\therefore$ $9a^2 + 12ab + 4b^2 = (3a + 2b)^2$

**EXAMPLE 14.7**

Factorise $16m^2 - 40m + 25$

Now $16m^2 = (4m)^2$,  $25 = -5)^2$  and  $-40m = 2 \times 4m \times (-5)$

$\therefore$ $16m^2 - 40m + 25 = (4m - 5)^2$

## The Factors of the Difference of Two Squares

It has previously been shown that

$$(a + b)(a - b) = a^2 - b^2$$

The factors of the difference of two squares are therefore the sum and the difference of the square roots of each of the given terms.

**EXAMPLE 14.8**

Factorise $9m^2 - 4n^2$

Now  $9m^2 = (3m)^2$,  and  $4n^2 = (2n)^2$

$\therefore$ $9m^2 - 4n^2 = (3m + 2n)(3m - 2n)$

**EXAMPLE 14.9**

Factorise $4x^2 - 9$

Now  $4x^2 = (2x)^2$  and  $9 = (3)^2$

$\therefore$ $4x^2 - 9 = (2x + 3)(2x - 3)$

## Exercise 14.2

Factorise:

1) $x^2 + 4x + 3$

2) $x^2 + 6x + 8$

3) $x^2 - 3x + 2$

4) $x^2 + 2x - 15$

5) $x^2 + 6x - 7$

6) $x^2 - 5x - 14$

7) $x^2 - 2xy - 3y^2$

8) $2x^2 + 13x + 15$

9) $3p^2 + p - 2$

10) $4x^2 - 10x - 6$

11) $3m^2 - 8m - 28$

12) $21x^2 + 37x + 10$

13) $10a^2 + 19a - 15$

14) $6x^2 + x - 35$

15) $6p^2 + 7pq - 3q^2$

16) $12x^2 - 5xy - 2y^2$

17) $x^2 + 2xy + y^2$

18) $4x^2 + 12x + 9$

19) $p^2 + 4pq + 4q^2$

20) $9x^2 + 6x + 1$

21) $m^2 - 2mn + n^2$

22) $25x^2 - 20x + 4$

23) $x^2 - 4x + 4$

24) $m^2 - n^2$

25) $4x^2 - y^2$

26) $9p^2 - 4q^2$

27) $x^2 - 1/9$

28) $1 - b^2$

29) $1/x^2 - 1/y^2$

30) $121p^2 - 64q^2$

## ROOTS OF AN EQUATION

If either of two factors has zero value, then their product is zero. Thus if either $M = 0$ or $N = 0$ then $M \times N = 0$

Now suppose that either $\qquad x = 1$ or $\qquad x = 2$

$\therefore$ rearranging gives either $\qquad x - 1 = 0$ or $x - 2 = 0$

Hence $\qquad (x - 1)(x - 2) = 0$

since either of the factors has zero value.

If we now multiply out the brackets of this equation we have

$$x^2 - 3x + 2 = 0$$

and we know that $x = 1$ and $x = 2$ are values of $x$ which satisfy this equation. The values 1 and 2 are called the solutions or *roots* of the equation $x^2 - 3x + 2 = 0$

### EXAMPLE 14.10

Find the equation whose roots are $-2$ and 4

From the values given either $\quad x = -2$ or $\qquad x = 4$

$\therefore \qquad$ either $\quad x + 2 = 0 \qquad$ or $\quad x - 4 = 0$

Hence $\qquad (x + 2)(x - 4) = 0$

since either of the factors has zero value.

$\therefore \quad$ Multiplying out gives $\quad x^2 - 2x - 8 = 0$

**EXAMPLE 14.11**

Find the equation whose roots are 3 and $-3$

From the values given either     $x = 3$   or     $x = -3$

$\therefore$                       either   $x - 3 = 0$   or   $x + 3 = 0$

Hence                     $(x - 3)(x + 3) = 0$

since either of the factors has zero value.

Multiplying out we have     $x^2 - 9 = 0$

**EXAMPLE 14.12**

Find the equation whose roots are 5 and 0.

From the given values given either     $x = 5$   or   $x = 0$

$\therefore$                       either   $x - 5 = 0$   or   $x = 0$

Hence                     $x(x - 5) = 0$

since either of the factors has zero value.

And multiplying out we have     $x^2 - 5x = 0$

## Exercise 14.3

Find the equations whose roots are:

1) 3, 1                          2) 2, $-4$

3) $-1$, $-2$                      4) 1.6, 0.7

5) 2.73, $-1.66$                  6) $-4.76$, $-2.56$

7) 0, 1.4                        8) $-4.36$, 0

9) $-3.5$, $+3.5$               10) repeated, each $= 4$

## QUADRATIC EQUATIONS

An equation of the type $ax^2 + bx + c = 0$, involving $x$ in the second degree and containing no higher power of $x$, is called a *quadratic equation*. The constants $a$, $b$ and $c$ have any numerical values. Thus,

$$x^2 - 9 = 0 \quad \text{where } a = 1, \ b = 0 \text{ and } c = -9$$

$$x^2 - 2x - 8 = 0 \quad \text{where } a = 1, \ b = -2 \text{ and } c = -8$$

$$2.5x^2 - 3.1x - 2 = 0 \quad \text{where } a = 2.5, \ b = -3.1 \text{ and } c = -2$$

are all examples of quadratic equations. A quadratic equation may contain only the square of the unknown quantity, as in the first of the above equations, or it may contain both the square and the first power as in the other two.

# 1. Solution by Factors

This method is the reverse of the procedure used to find an equation when given the roots. We shall now start with the equation and proceed to solve the equation and find the roots.

We shall again use the fact that if the product of two factors is zero then one factor or the other must be zero. Thus if $M \times N = 0$ then either $M = 0$ or $N = 0$

When the factors are easy to find the factor method is very quick and simple. However do not spend too long trying to find factors: if they are not easily found use the formula given in the next method (page 231) to solve the equation.

### EXAMPLE 14.13

Solve the equation  $(2x + 3)(x - 5) = 0$

Since the product of the two factors $2x + 3$ and $x - 5$ is zero then either

$$2x + 3 = 0 \quad \text{or} \quad x - 5 = 0$$

Hence
$$x = -\frac{3}{2} \quad \text{or} \quad x = 5$$

### EXAMPLE 14.14

Solve the equation     $6x^2 + x - 15 = 0$

Factorising gives  $(2x - 3)(3x + 5) = 0$

∴                either   $2x - 3 = 0$   or   $3x + 5 = 0$

Hence
$$x = \frac{3}{2} \quad \text{or} \quad x = -\frac{5}{3}$$

**EXAMPLE 14.15**

Solve the equation $14x^2 = 29x - 12$

Bring all the terms to the left-hand side:

$$14x^2 - 29x + 12 = 0$$

$\therefore$
$$(7x - 4)(2x - 3) = 0$$

$\therefore$ either $\quad 7x - 4 = 0 \quad$ or $\quad 2x - 3 = 0$

Hence $\qquad\qquad\qquad x = \dfrac{4}{7} \quad$ or $\quad x = \dfrac{3}{2}$

**EXAMPLE 14.16**

Find the roots of the equation $x^2 - 16 = 0$

Factorising gives $\qquad (x - 4)(x + 4) = 0$

$\therefore$ either $\quad x - 4 = 0 \quad$ or $\quad x + 4 = 0$

Hence $\qquad\qquad x = 4 \quad$ or $\quad x = -4$

In this case an alternative method may be used:

Rearranging the given equation gives $\quad x^2 = 16$

and taking the square root of both sides gives $\quad x = \sqrt{16} = \pm 4$

Remember that when we take a square root we must insert the $\pm$ sign, because $(+ 4)^2 = 16$ and $(- 4)^2 = 16$

**EXAMPLE 14.17**

Solve the equation $x^2 - 2x = 0$

Factorising gives $\qquad\qquad x(x - 2) = 0$

$\therefore$ either $\quad x = 0 \quad$ or $\quad x - 2 = 0$

Hence $\qquad\qquad x = 0 \quad$ or $\quad x = 2$

*Note*: The solution $x = 0$ must not be omitted as it is a solution in the same way as $x = 2$ is a solution. Equations should not be divided through by variables, such as $x$, since this removes a root of the equation.

**EXAMPLE 14.18**

Solve the equation $x^2 - 6x + 9 = 0$

Factorising gives $\qquad (x - 3)(x - 3) = 0$

$\therefore$        either   $x - 3 = 0$   or   $x - 3 = 0$

Hence               $x = 3$   or   $x = 3$

In this case there is only one arithmetical value for the solution. Technically, however, there are two roots and when they have the same numerical value they are said to be repeated roots.

## 2. Solution by Formula

In general, quadratic expressions do not factorise and therefore some other method of solving quadratic equations must be used.

Consider the expression    $ax^2 + bx = a\left(x^2 + \dfrac{b}{a}x\right)$

If we add (half the coefficient of $x$)$^2$ to the terms inside the bracket we get

$$ax^2 + bx = a\left[x^2 + \frac{b}{a}x + \left(\frac{b}{2a}\right)^2\right] - a\left(\frac{b}{2a}\right)^2$$

$$= a\left(x + \frac{b}{2a}\right)^2 - \frac{b^2}{4a}$$

We are said to have completed the square of $ax^2 + bx$.

We shall now establish a formula which may be used to solve any quadratic equation.

If                   $ax^2 + bx + c = 0$

then               $ax^2 - bx = -c$

Completing the square of the LHS gives

$$a\left(x + \frac{b}{2a}\right)^2 - \frac{b^2}{4a} = -c$$

$\therefore$           $4a^2\left(x + \dfrac{b}{2a}\right)^2 - b^2 = -4ac$

or            $4a^2\left(x + \dfrac{b}{2a}\right)^2 = b^2 - 4ac$

Taking the square root of both sides

$$2a\left(x + \frac{b}{2a}\right) = \pm\sqrt{b^2 - 4ac}$$

from which          $x = \dfrac{-b \pm \sqrt{b^2 - 4ac}}{2a}$

The *standard form* of the *quadratic equation* is:

$$ax^2 + bx + c = 0$$

As shown on the previous page the *solution* of this equation is:

$$x = \frac{-b \pm \sqrt{b^2 - 4ac}}{2a}$$

**EXAMPLE 14.19**

Solve the equation $3x^2 - 8x + 2 = 0$

Comparing with $ax^2 + bx + c = 0$, we have $a = 3$, $b = -8$ and $c = 2$

Substituting these values in the formula, we have

$$x = \frac{-(-8) \pm \sqrt{(-8)^2 - 4 \times 3 \times 2}}{2 \times 3}$$

$$= \frac{8 \pm \sqrt{64 - 24}}{6} = \frac{8 \pm \sqrt{40}}{6} = \frac{8 \pm 6.325}{6}$$

$\therefore$ either $\quad x = \dfrac{8 + 6.325}{6} \quad$ or $\quad x = \dfrac{8 - 6.325}{6}$

Hence $\quad x = 2.39 \qquad$ or $\quad x = 0.28$

It is important that we check the solutions in case we have made an error. We may do this by substituting the values obtained in the left-hand side of the given equation and checking that the solution is zero, or approximately zero.

Thus when $x = 2.39$ we have LHS $= 3(2.39)^2 - 8(2.39) + 2 \simeq 0$ and when $x = 0.28$ we have LHS $= 3(0.28)^2 - 8(0.28) + 2 \simeq 0$

**EXAMPLE 14.20**

Solve the equation $2.13x^2 + 0.75 - 6.89 = 0$

Here $a = 2.13$, $b = 0.75$, $c = -6.89$

$$x = \frac{-0.75 \pm \sqrt{(0.75)^2 - 4(2.13)(-6.89)}}{2 \times 2.13}$$

$$= \frac{-0.75 \pm \sqrt{0.5625 + 58.70}}{4.26} = \frac{-0.75 \pm \sqrt{59.26}}{4.26}$$

$$= \frac{-0.75 \pm 7.698}{4.26}$$

∴ either $x = \dfrac{-0.75 + 7.698}{4.26}$ or $x = \dfrac{-0.75 - 7.698}{4.26}$

Hence $x = 1.631$ or $x = -1.983$

*Solution check*

When $x = 1.631$

we have LHS $= 2.13(1.631)^2 + 0.75(1.631) - 6.89 \simeq 0$

When $x = -1.983$

we have LHS $= 2.13(-1.983)^2 + 0.75(-1.983) - 6.89 \simeq 0$

**EXAMPLE 14.21**

Solve the equation $x^2 + 4x + 5 = 0$

Here $a = 1$, $b = 4$ and $c = 5$

$$\therefore \quad x = \frac{-4 \pm \sqrt{4^2 - 4(1)(5)}}{2(1)} = \frac{-4 \pm \sqrt{16 - 20}}{2} = \frac{-4 \pm \sqrt{-4}}{2}$$

Now when a number is squared the answer must be a positive quantity because two quantities having the same sign are being multiplied together. Therefore the square root of a negative quantity, as $\sqrt{-4}$ in the above equation, has no arithmetical meaning and is called an imaginary quantity. The equation $x^2 + 4x + 5 = 0$ is said to have imaginary or complex roots. Equations which have complex roots are beyond the scope of this book and are dealt with in more advanced mathematics.

# Exercise 14.4

Solve the following equations by the factor method:

1) $x^2 - 36 = 0$
2) $4x^2 - 6.25 = 0$
3) $9x^2 - 16 = 0$
4) $x^2 + 9x + 20 = 0$
5) $x^2 + x - 72 = 0$
6) $3x^2 - 7x + 2 = 0$
7) $m^2 = 6m - 9$
8) $m^2 + 4m + 4 = 36$
9) $14q^2 = 29q - 12$
10) $9x + 28 = 9x^2$

Solve the following equations by using the quadratic formula:

11) $4x^2 - 3x - 2 = 0$      12) $x^2 - x + \frac{1}{4} = \frac{1}{9}$

13) $3x^2 + 7x - 5 = 0$      14) $7x^2 + 8x - 2 = 0$

15) $5x^2 - 4x - 1 = 0$      16) $2x^2 - 7x = 3$

17) $x^2 + 0.3x - 1.2 = 0$      18) $2x^2 - 5.3x + 1.25 = 0$

Solve the following equations:

19) $x(x + 4) + 2x(x + 3) = 5$      20) $x^2 - 2x(x - 3) = -20$

21) $\dfrac{2}{x + 2} + \dfrac{3}{x + 1} = 5$      22) $\dfrac{x + 2}{3} - \dfrac{5}{x + 2} = 4$

23) $\dfrac{6}{x} - 2x = 2$      24) $40 = \dfrac{x^2}{80} + 4$

25) $\dfrac{x + 2}{x - 2} = x - 3$      26) $\dfrac{1}{x + 1} - \dfrac{1}{x + 3} = 15$

# PROBLEMS INVOLVING QUADRATIC EQUATIONS

### EXAMPLE 14.22

The distance, $s$ m, moved by a vehicle in time, $t$ s, with an initial velocity, $v_1$ m/s, and a constant acceleration, $a$ m/s$^2$, is given by $s = v_1 t + \frac{1}{2}at^2$. Find the time taken to cover 84 m with a constant acceleration 2 m/s$^2$ if the initial velocity is 5 m/s.

Using $s = 84$ m, $v_1 = 5$ m/s and $a = 2$ m/s$^2$ we have

$$84 = 5t + \tfrac{1}{2}\,2t^2$$

from which $\qquad\qquad t^2 + 5t - 84 = 0$

Factorising gives $\qquad (t + 12)(t - 7) = 0$

∴ $\qquad\qquad$ either $\quad t + 12 = 0 \qquad$ or $\quad t - 7 = 0$

∴ $\qquad\qquad$ either $\qquad\quad t = -12 \quad$ or $\qquad t = 7$

Now the solution $t = -12$ is not acceptable since negative time has no meaning in this question. Thus the required time is 7 seconds.

*Solution check*

When $t = 7$ we have LHS $= 7^2 + 5 \times 7 - 84 = 0$

**EXAMPLE 14.23**

The diagonal of a rectangle is 15 m long and one side is 2 m longer than the other. Find the dimensions of the rectangle.

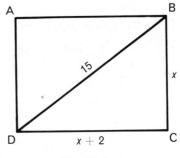

In Fig. 14.1, let the length of BC be $x$ cm. The length of CD is then $(x + 2)$ m. $\triangle BCD$ is right-angled and so by Pythagoras,

$$x^2 + (x + 2)^2 = 15^2$$

Fig. 14.1

$$\therefore \quad x^2 + x^2 + 4x + 4 = 225$$

$$\therefore \quad 2x^2 + 4x - 221 = 0$$

Here $a = 2$, $b = 4$ and $c = -221$

$$\therefore \quad x = \frac{-4 \pm \sqrt{4^2 - 4 \times 2 \times (-221)}}{2 \times 2}$$

$$\therefore \quad x = 9.56 \quad \text{or} \quad -11.56$$

Since the answer cannot be negative, then $x = 9.56$ m

Now $\qquad\qquad x + 2 = 11.56$ m

$\therefore$ the rectangle has adjacent sides equal to 9.56 m and 11.56 m.

*Solution check*

When $x = 9.56$ we have LHS $= 2(9.56)^2 + 4(9.56) - 221 \approx 0$

**EXAMPLE 14.24**

A section of an air duct is shown by the full lines in Fig. 14.2.

a) Show that: $w^2 - 2Rw + \dfrac{R^2}{4} = 0$

b) Find the value of $w$ when $R = 2$ m

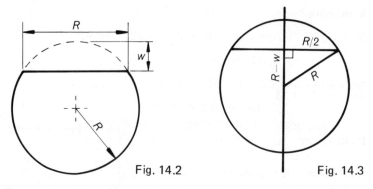

Fig. 14.2                          Fig. 14.3

Using the construction shown in Fig. 14.3 we have, by Pythagoras,

then
$$(R-w)^2 + \left(\frac{R}{2}\right)^2 = R^2$$

$\therefore$
$$R^2 - 2Rw + w^2 + \frac{R^2}{4} = R^2$$

$\therefore$
$$w^2 - 2Rw + \frac{R^2}{4} = 0$$

When $R = 2$,
$$w^2 - 4w + 1 = 0$$

Here $a = 1$, $b = -4$ and $c = 1$

$\therefore$
$$w = \frac{-(-4) \pm \sqrt{(-4)^2 - 4 \times 1 \times 1}}{2 \times 1}$$

Hence    $w = 3.732$   or   $0.268\,\text{m}$

Now $w$ must be less than $2\,\text{m}$, thus $w = 0.268\,\text{m}$

*Solution check*

When $w = 0.268$  we have  LHS $= (0.268)^2 - 4(0.268) + 1 \simeq 0$

### EXAMPLE 14.25

Specified quantities of hydrogen and iodine are mixed together in a vessel of given volume at a certain temperature. If the equilibrium constant, $K_c$, is given by $K_c = \dfrac{(2-x)(1-x)}{(2x)^2}$ find the quantity, $(2-x)$, of hydrogen at equilibrium, given that $K_c = 0.018$

If we substitute $K_c = 0.018$ in the given expression for $K_c$

then
$$0.018 = \frac{(2-x)(1-x)}{(2x)^2}$$

from which  $\qquad 0.018(2x)^2 = (2-x)(1-x)$

∴  $\qquad\qquad\qquad 0.072x^2 = 2-3x+x^2$

∴  $\qquad 0.928x^2-3x+2 = 0$

Now the standard form of a quadratic equation is

$$ax^2+bx+c = 0$$

from which  $\qquad\qquad x = \dfrac{-b\pm\sqrt{b^2-4ac}}{2a}$

Comparing the quadratic equations we have $a = 0.928$, $b = -3$ and $c = 2$

Thus  $\qquad\qquad x = \dfrac{-(-3)\pm\sqrt{(-3)^2-4(0.928)2}}{2(0.928)}$

from which  $\quad x = 2.293 \quad$ or $\quad 0.940$

Now if $x = 2.293$ then the quantity of hydrogen is $(2-2.293)$ $= -0.293$ which is unacceptable since there cannot be a negative quantity. We therefore reject this answer and use the other.

Hence the required quantity of hydrogen is $(2-0.94) = 1.06$

*Solution check*

In this case we would check the solution by substituting $x = 0.940$ into the right-hand side of the given expression and hope that we obtain 0.018

Thus  $\qquad$ RHS $= \dfrac{(2-0.940)(1-0.940)}{(2\times0.940)^2} = 0.018$

# Exercise 14.5

1) The length $L$ of a wire stretched tightly between two supports in the same horizontal line is given by

$$L = S+\frac{8D^2}{3S}$$

where $S$ is the span and $D$ is the (small) sag. If $L = 150$ and $D = 5$, find the value of $S$.

2) In a right-angled triangle the hypotenuse is twice as long as one of the sides forming the right angle. The remaining side is 80 mm long. Find the length of the hypotenuse.

3) The area of a rectangle is $61.75\,m^2$. If the length is 3 m more than the width find the dimensions of the rectangle.

4) The total surface area of a cylinder whose height is 75 mm is 29 000 mm$^2$. Find the radius of the cylinder.

5) If a segment of a circle has a radius $R$, a height $H$ and a length of chord $W$ show that

$$R = \frac{W^2}{8H} + \frac{H}{2}$$

Rearrange this equation to give a quadratic equation for $H$ and hence find $H$ when $R = 12$ m and $W = 8$ m.

6) Fig. 14.4 shows a template whose area is 9690 mm$^2$. Find the value of $r$.

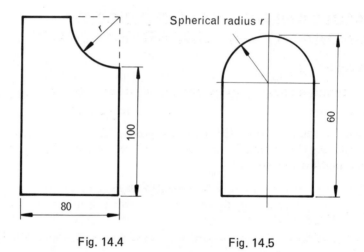

Fig. 14.4                    Fig. 14.5

7) A pressure vessel is of the shape shown in Fig. 14.5, the radius of the vessel being $r$ mm. If the surface area is 30 000 mm$^3$ find $r$.

8) The total iron loss in a transformer is given by the equation $P = 0.1f + 0.006f^2$. If $P = 20$ watts, find the value of the frequency $f$.

9) The volume of a frustum of a cone is given by the formula $V = \frac{1}{3}\pi h(R^2 + rR + r^2)$ where $h$ is the height of the frustum and $R$ and $r$ are the radii at the large and small ends respectively. If $h = 9$ m, $R = 4$ m and the volume is 337.2 m$^3$, what is the value of $r$?

10) A square steel plate is pierced by a square tool leaving a margin of 20 mm all round. The area of the hole is one third that of the original plate. What are the dimensions of the original plate?

**11)** The velocity, $v$, of a body in terms of time, $t$, is given by the expression $v = 3t^2 - 6t - 3$. Find the times at which the velocity is zero.

**12)** The value of the equilibrium constant, $K_c$, when phosphorus pentachloride vapour is heated is given by the expression $K_c = \dfrac{x^2}{V(1-x)}$ where, for an equilibrium mixture, $V$ is the total volume and $x$ is the quantity of chlorine. Find the value of $x$ if $K_c = 0.012$ and $V = 24$

# SIMULTANEOUS SOLUTION OF A QUADRATIC AND LINEAR EQUATION

### EXAMPLE 14.26

Solve simultaneously the equations:     $y = x^2 + 3x - 4$      [1]

              and   $y = 2x + 4$      [2]

Substituting the value of $y$ given by equation [2], that is $y = 2x + 4$ into equation [1] we have

$$2x + 4 = x^2 + 3x - 4$$

from which      $x^2 + x - 8 = 0$

This equation does not factorise so we will use the formula. Here $a = 1$, $b = 1$ and $c = -8$

$$\therefore \qquad x = \frac{-1 \pm \sqrt{1^2 - 4(1)(-8)}}{2(1)} = \frac{-1 \pm \sqrt{33}}{2}$$

$$\therefore \quad \text{either} \quad x = \frac{-1 + 5.745}{2} \quad \text{or} \quad x = \frac{-1 - 5.745}{2}$$

$\therefore$     either    $x = 2.372$       or    $x = -3.372$

Now for each of these values of $x$ there will be a corresponding value of $y$. We may find these values of $y$ by substituting the values $x = 2.372$ and $x = -3.372$ into either of the given equations.

Therefore from equation [2]

    when $x = 2.372$,     $y = 2(2.372) + 4 = 8.744$

and

    when $x = -3.372$,    $y = 2(-3.372) + 4 = -2.744$

Hence the required solutions are

| $x$ | 2.372 | $-3.372$ |
|---|---|---|
| $y$ | 8.744 | $-2.744$ |

These solutions may be checked by substituting the values into equation [1]:

RHS $= (2.372)^2 + 3(2.372) - 4 = 8.742 \simeq$ LHS

and

RHS $= (-3.372)^2 + 3(-3.372) - 4 = -2.745 \simeq$ LHS

The inaccuracies occur because the values have been rounded off correct to the third place of decimals.

### EXAMPLE 14.27

Solve simultaneously the equations:    $y^2 - 2y - 4 = x$    [1]

and   $x + 3y - 2 = 0$    [2]

Now equation [2] may be rearranged to give $x = -3y + 2$ and if we substitute this value of $x$ into equation [1] we have

$$y^2 - 2y - 4 = -3y + 2$$

from which    $y^2 + y - 6 = 0$

and factorising    $(y + 3)(y - 2) = 0$    gives

Thus    either $y + 3 = 0$    or $y - 2 = 0$

∴    either    $y = -3$    or    $y = 2$

Now for each of these values of $y$ there will be a corresponding value of $x$. We may find these values of $x$ by substituting the values of $y$ into either of the given equations. Therefore from equation [2]

when $y = -3$, $x + 3(-3) - 2 = 0$ ∴ $x = 11$

and when $y = 2$, $x + 3(2) - 2 = 0$ ∴ $x = -4$

Hence the required solutions are

| $x$ | 11 | $-4$ |
|---|---|---|
| $y$ | $-3$ | 2 |

These solutions may be checked by substituting the values into equation [1]:

LHS $= (-3)^2 - 2(-3) - 4 = 9 + 6 - 4 = 11 =$ RHS

and LHS $= (2)^2 - 2(2) - 4 = 4 - 4 - 4 = -4 =$ RHS

# Exercise 14.6

Solve simultaneously, by substitution:

1) $y = 3x^2 - 3x - 1$
   $y = 4x - 3$

2) $y = 8x^2 - 2$
   $y = 1 - 5x$

3) $y = 4x^2 - 2x - 1$
   $x - y + 1 = 0$

4) $x = y^2 - 0.4y - 1.5$
   $x + 0.7y + 0.3 = 0$

5) $y = 7x^2 + 6x - 3$
   $2x + y = -1$

6) $y = x^2 + 1.213x + 0.574$
   $y = 2.213x + 0.435$

7) A cutting is in the shape of an isosceles trapezium which is $2x$ metres wide at the top, $2y$ metres wide at the bottom and has a vertical height of 8 m. The cross-sectional area of the cutting is $144\,m^2$ and its perimeter (two sloping sides and the base) is 32 m. It can be shown that

$$x + y = 18 \qquad\qquad [1]$$

and $\qquad\qquad x^2 - 2xy + 4y = -60 \qquad\qquad [2]$

Solve these equations and hence find the bottom width of the cross-section of the cutting.

8) Fig. 14.6 shows a hot water cylinder with a surface area of $138\,m^2$. Show that

$$3\pi r^2 + 2\pi rh = 138 \qquad\qquad [1]$$

and $\qquad\qquad r + h = 10 \qquad\qquad [2]$

By solving this pair of simultaneous equations find the values of $r$ and $h$.

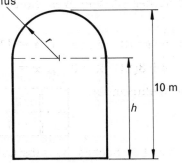

Spherical radius

10 m

$h$

Fig. 14.6

# LOGARITHMS AND EXPONENTIALS

## LOGARITHMS

If $N$ is a number such that

we may write this in the alternative form

$$N = b^x$$
$$\log_b N = x$$

which, in words, is    'the logarithm of $N$, to the base $b$, is $x$'

or    '$x$ is the logarithm of $N$ to the base $b$'

The word 'logarithm' is often abbreviated to just 'log'.

It is helpful to remember that:

$$\text{Number} = \text{Base}^{\text{logarithm}}$$

Alternatively in words:

The log of a number is the power to which the base must be raised to give that number

Thus:

| We may write | $8 = 2^3$ | We may write | $81 = 3^4$ |
|---|---|---|---|
| in log form as | $\log_2 8 = 3$ | in log form as | $\log_3 81 = 4$ |

| We may write | $2 = \sqrt{4}$ | We may write | $\dfrac{1}{4} = \dfrac{1}{2^2}$ |
|---|---|---|---|
| or | $2 = 4^{1/2}$ | | |
| or | $2 = 4^{0.5}$ | or | $0.25 = 2^{-2}$ |
| in log form as | $\log_4 2 = 0.5$ | in log form as | $\log_2 0.25 = -2$ |

242

**EXAMPLE 15.1**

If $\log_7 49 = x$, find the value of $x$.

Writing the equation in index form we have    $49 = 7^x$

$$\text{or} \quad 7^2 = 7^x$$

Since the bases are the same on both sides of the equation the indices must be the same.    Thus    $x = 2$

**EXAMPLE 15.2**

If $\log_x 8 = 3$, find the value of $x$.

Writing this equation in index form we have    $8 = x^3$

$$\text{or} \quad 2^3 = x^3$$

Since the indices on both sides of the equation are the same the bases must be the same.    Thus    $x = 2$

# THE VALUE OF $\log_b 1$

Let    $\log_b 1 = x$

then in index form    $1 = b^x$

Now the only value of the index $x$ which will satisfy this expression is zero.

Hence    $\log_b 1 = 0$

Thus:    | To any base the value of log 1 is zero |

# THE VALUE OF $\log_b b$

Let    $\log_b b = x$

then in index form    $b = b^x$

Now the only value of the index $x$ which will satisfy the expression is unity.

Hence    $\log_b b = 1$

Thus:    | The value of the log of a number to the same base is unity |

# THE VALUE OF $\log_b 0$

Let $$\log_b 0 = x$$

then in index form $$0 = b^x$$

Now consider, for example, the value of $10^x$:

If $x = -2$ then the value of $10^x$ will be $10^{-2} = \dfrac{1}{10^2}$

If $x = -20$ then the value of $10^x$ will be $10^{-20} = \dfrac{1}{10^{20}}$

If $x = -200$ then the value of $10^x$ will be $10^{-200} = \dfrac{1}{10^{200}}$

Now $\dfrac{1}{10^{200}}$ is a very small number indeed and from this pattern we may deduce that if the value of the index $x$ is an infinitely large negative number (called 'minus infinity', written as $-\infty$) then the value of $10^x$ would be zero. It follows that $b^{-\infty} = 0$

Hence $$\log_b 0 = -\infty$$

Thus:

> To any base the log of zero is minus infinity

# THE VALUE OF $\log_b (- N)$

Let $$\log_b(-N) = x$$

then in index form $$-N = b^x$$

If we examine this expression we can see that whatever the value of the negative number $N$, or whatever the value of the base, $b$, it is not possible to find a value for the index $x$ which will satisfy the expression.

Hence $\qquad \log_b(-N)$ has no real value

Thus

> Only positive numbers have real logarithms

## Exercise 15.1

Express in logarithmic form:

1) $n = a^x$ 　　　　　 2) $2^3 = 8$ 　　　　　 3) $5^{-2} = 0.04$

4) $10^{-3} = 0.001$   5) $x^0 = 1$   6) $10^1 = 10$

7) $a^1 = a$   8) $e^2 = 7.39$   9) $10^0 = 1$

Find the value of $x$ in each of the following:

10) $\log_x 9 = 2$   11) $\log_x 81 = 4$   12) $\log_2 16 = x$

13) $\log_5 125 = x$   14) $\log_3 x = 2$   15) $\log_4 x = 3$

16) $\log_{10} x = 2$   17) $\log_7 x = 0$   18) $\log_x 8 = 3$

19) $\log_x 27 = 3$   20) $\log_9 3 = x$   21) $\log_n n = x$

# LAWS OF LOGARITHMS

Let   $\log_b M = x$   and   $\log_b N = y$

or in index form   $M = b^x$   and   $N = b^y$

**(1)** Now   $MN = b^x \times b^y$

∴   $MN = b^{x+y}$

or in log form   $\log_b MN = x + y$

Hence   $\boxed{\log_b MN = \log_b M + \log_b N}$

In words this relationship is:

> The logarithm of two numbers multiplied together may be found by adding their individual logarithms

**(2)** Now   $\dfrac{M}{N} = \dfrac{b^x}{b^y}$

∴   $\dfrac{M}{N} = b^{x-y}$

or in log form   $\log_b \dfrac{M}{N} = x - y$

Hence   $\boxed{\log_b \dfrac{M}{N} = \log_b M - \log_b N}$

In words this relationship is:

> The logarithm of two numbers divided may be found by subtracting their individual logarithms

(3)  Now $$M^n = (b^x)^n$$

∴ $$M^n = b^{nx}$$

or in log form $$\log_b M^n = nx$$

Hence $$\boxed{\log_b M^n = n(\log_b M)}$$

In words this relationship is:

> The logarithm of a number raised to a power may be found by multiplying the logarithm of the number by the power

# LOGARITHMS TO THE BASE 10

Logarithms to the base 10 are called common logarithms and stated as $\log_{10}$ (or lg). When logarithmic tables are used to solve numerical problems, tables to this base are preferred as they are simpler to use than tables to any other base. Common logarithms are also used for scales on logarithmic graph paper and also for calculations on the measurement of sound.

# LOGARITHMS TO THE BASE 'e'

In higher mathematics all logarithms are taken to the base e, where e = 2.718 28. Logarithms to this base are often called natural logarithms. They are also called Naperian or hyperbolic logarithms.

Natural logarithms are stated as $\log_e$ (or ln).

# CHOICE OF BASE FOR CALCULATIONS

If an electronic calculating machine of the scientific type is used, then it is just as easy to use logarithms to the base e. Some machines have keys for both $\log_e$ and $\log_{10}$ but on the more limited models only $\log_e$ is given.

The natural logarithms is found by using the $\log_e$ (or ln) key and the natural antilogarithm is found by using the $e^x$ key.

The common logarithm is found using the $\log_{10}$ (or lg) key and the common antilogarithm is found using the $10^x$ key.

In the example which follows it is appreciated that use of the power key $x^y$ will give an immediate solution, but it is instructive to work through the alternative method of solution using logarithms.

**EXAMPLE 15.3**

Evaluate $3.714^{2.87}$.

Let
$$x = 3.714^{2.87}$$

and taking logarithms of both sides we have

$$\log x = \log 3.714^{2.87}$$

∴
$$\log x = 2.87 \times \log 3.714$$

∴
$$x = \text{antilog}\,(2.87 \times \log 3.714)$$

The base of the logarithms has not yet been chosen — the sequence given will be true for any base value.

Using a calculator with natural logarithms the sequence of operations is:

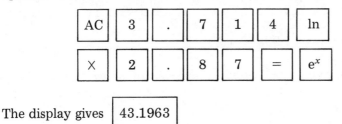

The display gives $\boxed{43.1963}$

Thus the answer is 43.2 correct to three significant figures.

# INDICIAL EQUATIONS

These are equations in which the number to be found is an index, or part of an index.

The method of solution is to reduce the given equation to an equation involving logarithms, as the following examples will illustrate.

**EXAMPLE 15.4**

If $8.79^x = 67.65$ find the value of $x$.

Now taking logarithms of both sides of the given equation we have

$$\log 8.79^x = \log 67.65$$

$$\therefore \qquad x(\log 8.79) = \log 67.65$$

$$\therefore \qquad x = \frac{\log 67.65}{\log 8.79}$$

The base of the logarithms has not yet been chosen, the above procedure being true for any base value.

The quickest way, that is without using the reciprocal $\boxed{\dfrac{1}{x}}$ key, is to find the value of the bottom line and put this into the memory. Then find the value of the top line and divide this by the content of the memory.

The sequence, using natural logarithms, would be:

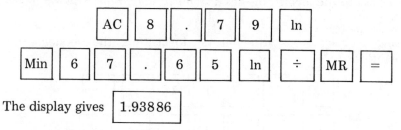

The display gives $\boxed{1.938\,86}$

Thus the answer is 1.94 correct to three significant figures.

**EXAMPLE 15.5**

Find the value of $x$ if $1.793^{(x+3)} = 20^{0.982}$

Now taking logarithms of both sides of the given equation we have

$$\log 1.793^{x+3} = \log 20^{0.982}$$

$$\therefore \qquad (x+3)(\log 1.793) = (0.982)(\log 20)$$

$$\therefore \qquad x+3 = \frac{(0.982)(\log 20)}{\log 1.793}$$

$$\therefore \qquad x = \frac{(0.982)(\log 20)}{\log 1.793} - 3$$

The procedure will be similar to that used in Example 15.4. The sequence of operations, using natural logarithms, is then:

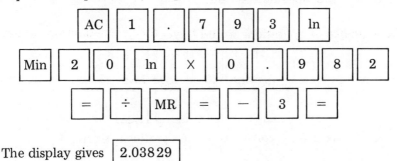

The display gives   2.038 29

Thus the answer is 2.04 correct to three significant figures.

## Exercise 15.2

Evaluate the following:

1) $11.57^{0.3}$        2) $15.62^{2.15}$        3) $0.6327^{0.5}$

4) $0.06521^{3.16}$        5) $27.15^{-0.4}$

Find the value of $x$ in the following:

6) $3.6^x = 9.7$        7) $0.9^x = 2.176$

8) $\left(\dfrac{1}{7.2}\right)^x = 1.89$        9) $1.4^{(x+2)} = 9.3$

10) $21.9^{(3-x)} = 7.334$        11) $2.79^{(x-1)} = 4.377^x$

12) $\left(\dfrac{1}{0.64}\right)^{(2+x)} = 1.543^{(x+1)}$        13) $\dfrac{1}{0.9^{(x-2)}} = 8.45$

## CALCULATIONS INVOLVING THE EXPONENTIAL FUNCTIONS, $e^x$ and $e^{-x}$

### EXAMPLE 15.6

Evaluate $50\,e^{2.16}$

The sequence of operations is:

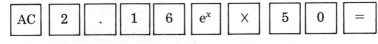

giving an answer 434 correct to three significant figures.

**EXAMPLE 15.7**

Evaluate $200\,e^{-1.34}$

The sequence of operations would then be:

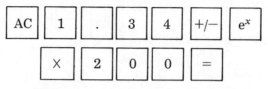

giving an answer 52.4 correct to three significant figures.


**EXAMPLE 15.8**

In a capacitive circuit the instantaneous voltage across the capacitor is given by $v = V(1-e^{-t/CR})$ where $V$ is the initial supply voltage, $R$ ohms the resistance, $C$ farads the capacitance, and $t$ seconds the time from the instant of connecting the supply voltage.

If $V = 200$, $R = 10000$, and $C = 20 \times 10^{-6}$ find the time when the voltage $v$ is 100 volts.

Substituting the given values in the equation we have

$$100 = 200(1-e^{t/\,20 \times 10^{-6} \times 10\,000})$$

$$\therefore \qquad \frac{100}{200} = 1-e^{-t/0.2}$$

$$\therefore \qquad 0.5 = 1-e^{-5t}$$

$$\therefore \qquad e^{-5t} = 1-0.5$$

$$\therefore \qquad e^{-5t} = 0.5$$

Thus in log form

$$\log_e 0.5 = -5t$$

$$\therefore \qquad t = -\frac{\log_e 0.5}{5}$$

The sequence of operation is:

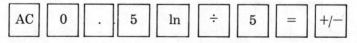

giving an answer 0.139 seconds correct to three significant figures.

**EXAMPLE 15.9**

$$R = \frac{(0.42)S}{l} \times \log_e \frac{d_2}{d_1}$$

refers to the insulation resistance of a wire. Find the value of $R$ when $S = 2000$, $l = 120$, $d_1 = 0.2$ and $d_2 = 0.3$

Substituting the given values gives

$$R = \frac{0.42 \times 2000}{120} \times \log_e \frac{0.3}{0.2}$$

$$= \frac{0.42 \times 2000}{120} \times \log_e 1.5$$

The sequence of operations would be:

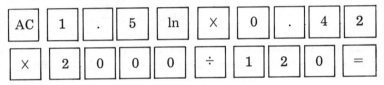

giving an answer 2.84 correct to three significant figures.

# Exercise 15.3

1) Find the numbers whose natural logarithms are:

(a) 2.76          (b) 0.677          (c) 0.09
(d) −3.46         (e) −0.543         (f) −0.078

2) Find the values of:

(a) $70 \, e^{2.5}$          (b) $150 \, e^{-1.34}$          (c) $3.4 \, e^{-0.445}$

3) The formula $L = 0.000644 \left( \log_e \dfrac{d}{r} + \dfrac{1}{4} \right)$ is used for calculating the self-inductance of parallel conductors. Find $L$ when $d = 50$ and $r = 0.25$

4) The inductance ($L$ microhenrys) of a straight aerial is given by the formula: $L = \dfrac{1}{500} \left( \log_e \dfrac{4l}{d} - 1 \right)$ where $l$ is the length of the aerial in mm and $d$ its diameter in mm. Calculate the inductance of an aerial 5000 m long and 2 mm in diameter.

5) Find the value of $\log_e \left( \dfrac{c_1}{c_2} \right)^2$ when $c_1 = 4.7$ and $c_2 = 3.5$

6) If $T = R \log_e \left( \dfrac{a}{a-b} \right)$ find $T$ when $R = 28$, $a = 5$ and $b = 3$.

7) When a chain of length $2l$ is suspended from two points $2d$ apart on the same horizontal level, $d = c \log_e \left( \dfrac{l + \sqrt{l^2 + c^2}}{c} \right)$. If $c = 80$ and $l = 200$ find $d$.

8) The instantaneous value of the current when an inductive circuit is discharging is given by the formula $i = Ie^{-Rt/L}$. Find the value of this current, $i$, when $I = 6$, $R = 30$, $L = 0.5$ and $t = 0.005$

9) In a circuit in which a resistor is connected in series with a capacitor the instantaneous voltage across the capacitor is given by the formula $v = V(1 - e^{-t/CR})$. Find this voltage, $v$, when $V = 200$, $C = 40 \times 10^{-6}$, $R = 100000$ and $t = 1$

10) In the formula $v = Ve^{-Rt/L}$ the values of $v$, $V$, $R$ and $L$ are 50, 150, 60 and 0.3 respectively. Find the corresponding value of $t$.

11) The instantaneous charge in a capacitive circuit is given by $q = Q(1 - e^{-t/CR})$. Find the value of $t$ when $q = 0.01$, $Q = 0.015$, $C = 0.0001$, and $R = 7000$

# EXPONENTIAL GRAPHS

Curves which have equations of the type $e^x$ and $e^{-x}$ are called exponential graphs.

We may plot the graphs of $e^x$ and $e^{-x}$ by using mathematical tables to find values of $e^x$ and $e^{-x}$ for chosen values of $x$. We should remember that any number to a zero power is unity: hence $e^0 = 1$

Drawing up a table of values we have:

| $x$ | $-2$ | $-1$ | 0 | 1 | 2 |
|------|------|------|------|------|------|
| $e^x$ | 0.14 | 0.37 | 1 | 2.72 | 7.39 |
| $-x$ | 2 | 1 | 0 | $-1$ | $-2$ |
| $e^{-x}$ | 7.39 | 2.72 | 1 | 0.37 | 0.14 |

For convenience both the curves are shown plotted on the same axes in Fig. 15.1. Although the range of values chosen for $x$ is limited the overall shape of the curves is clearly shown.

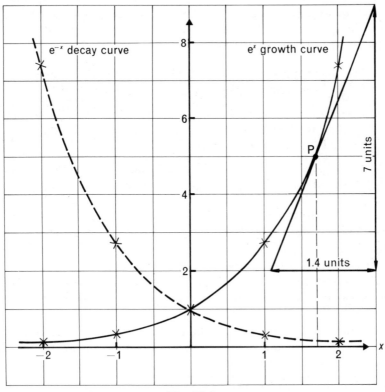

Fig. 15.1

The rate at which a curve is changing at any point is given by the gradient of the tangent at that point.

Remember the sign convention for gradients is

Positive gradient          Negative gradient

Now the gradient at any point on the $e^x$ graph is positive and so the rate of change is positive. In addition the rate of change increases as the values of $x$ increase. A graph of this type is called a *growth curve*.

The gradient at any point on the $e^{-x}$ graph is negative and so the rate of change is negative. In addition the rate of change decreases as the values of $x$ decrease. A graph of this type is called a *decay curve*.

# AN IMPORTANT PROPERTY OF THE EXPONENTIAL FUNCTION, $e^x$

Suppose we choose any point, P, on the graph of the function $e^x$ shown in Fig. 15.1 and draw a tangent to the curve at P. We may find the gradient of the curve at P by finding the gradient of the tangent using the right-angled triangle shown. The gradient at $P = \dfrac{7}{1.4} = 5$. Now this is also the value of $e^x$ at P. The reader may like to draw the curve of $e^x$ and check that the gradient at various points is always equal to the value of $e^x$ at corresponding points. This illustrates the important, and unique, property of the exponential function $e^x$ which is:

> At any point on the curve of the exponential function $e^x$ the gradient of the curve is equal to the value of the function

### EXAMPLE 15.10

The population size, $N$, at a certain time, $t$ hours, after commencement of growth of a unit sized population is given by the exponential growth relationship $N = e^{0.8t}$. Show that the instantaneous rate of growth is proportional to the population size.

The values of $N$ for corresponding values of $t$ may be found using the scientific calculator. The table of values shows results for values of $t$ from zero to 4 hours:

| $t$ hours | 0 | 0.5 | 1 | 1.5 | 2 | 2.5 | 3 | 3.5 | 4 |
|---|---|---|---|---|---|---|---|---|---|
| $N = e^{0.8t}$ | 1 | 1.49 | 2.23 | 3.32 | 4.95 | 7.39 | 11.0 | 16.4 | 24.5 |

The curve is shown plotted in Fig. 15.2.

The instantaneous rate of growth means the rate of growth at any instant. This is given by the gradient of the curve at any particular point. The gradient may be found by drawing a tangent to the curve at the point and calculating its slope by constructing a suitable right-angled triangle.

At P, using the right-angled triangle shown, the gradient $= \dfrac{5.4}{3} = 1.8$

Also the value of $N$ is 2.23.

Hence        the ratio:    $\dfrac{\text{Gradient}}{N \text{ value}} = \dfrac{1.8}{2.23} = 0.81$

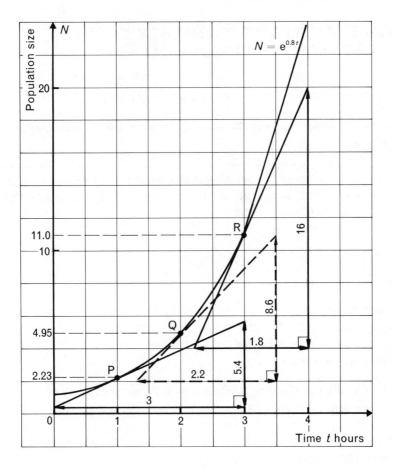

Fig. 15.2

Similarly at Q the ratio: $\dfrac{\text{Gradient}}{N \text{ value}} = \dfrac{8.6/2.2}{4.95} = 0.79$

And at $R$    the ratio: $\dfrac{\text{Gradient}}{N \text{ value}} = \dfrac{16/1.8}{11} = 0.81$

The reader may like to plot the curve and calculate the ratio $\dfrac{\text{Gradient}}{N \text{ value}}$ for other points. We can see from these results that the value of the ratio (within the limitations of accuracy of values obtained from the graph) is constant—in this case 0.80

It is, therefore, reasonable to assume that at any point on the exponential curve the ratio:

$$\frac{\text{Gradient}}{N \text{ value}} = \text{Constant}$$

or when rearranged     $\text{Gradient} = (\text{Constant})(N \text{ value})$

∴     $\text{Gradient} \propto N \text{ value}$

Thus the instantaneous rate of growth is proportional to the population size.

> From this we may conclude that a property of an exponential curve is that, at any point, the gradient is proportional to the $N$ value (i.e. the ordinate)

## EXAMPLE 15.11

The instantaneous e.m.f. in an inductive circuit is given by the expression $100e^{-4t}$ volts, where $t$ is time in seconds. Plot the graph of the e.m.f. for values of $t$ from 0 to 0.5 seconds, and use the graph to find:

a)  the value of the e.m.f. when $t = 0.25$ seconds, and
b)  the rate of change of the e.m.f. when $t = 0.1$ seconds.

The graph is shown plotted in Fig. 15.3, from values obtained from the sequence of operations:

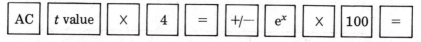

| AC | $t$ value | X | 4 | = | +/− | $e^x$ | X | 100 | = |

a)  The point P on the curve is at 0.25 seconds shown on the $t$ scale and the corresponding value of e.m.f. can be read directly from the vertical axis scale. The value is 37 volts.

b)  The point Q on the graph is at 0.1 seconds. Now the rate of change of the curve at Q is given by the gradient of the tangent at Q. This gradient may be found by constructing a suitable right-angled triangle such as MNO in Fig. 15.3, and finding the ratio $\dfrac{\text{MO}}{\text{ON}}$. Hence the

$$\text{Gradient at } Q = \frac{\text{MO}}{\text{ON}} = \frac{94 \text{ volts}}{0.35 \text{ seconds}} = 269 \text{ volts per second}$$

According to the sign convention a line sloping downwards from left to right has a negative gradient.

Hence the gradient at Q is $-269$ volts per second, which means that the rate of change of the curve at Q is $-269$ volts per second.

This is the same as saying that the e.m.f. at $t = 0.1$ seconds is decreasing at the rate of 269 volts per second.

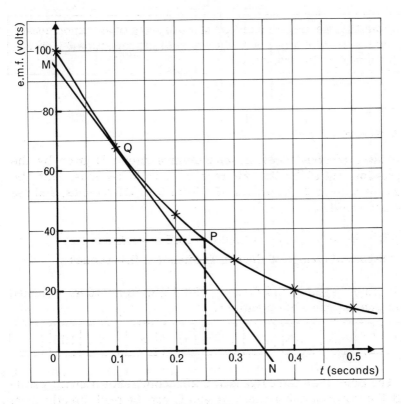

Fig. 15.3

**EXAMPLE 15.12**

The formula $i = 2(1-e^{-10t})$ gives the relationship between the instantaneous current $i$ amperes and the time $t$ seconds in an inductive circuit. Plot a graph of $i$ against $t$ taking values of $t$ from 0 to 0.3 seconds at intervals of 0.05 seconds. Hence find:

a) the initial rate of growth of the current $i$ when $t = 0$, and
b) the time taken for the current to increase from 1 to 1.6 amperes.

The curve is shown plotted in Fig. 15.4 from values obtained from the sequence of operations:

| AC | $t$ value | $\times$ | 1 | 0 | $=$ | +/− | $e^x$ | +/− |

| $+$ | 1 | $\times$ | 2 | $=$ |

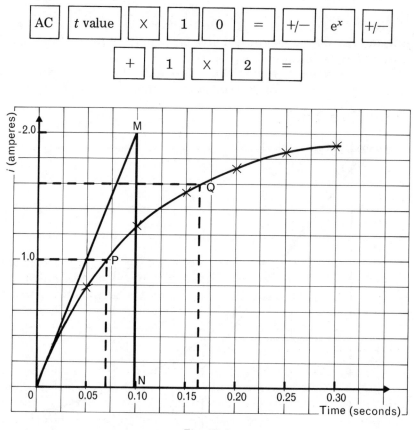

Fig. 15.4

a) When $t = 0$ the initial rate of growth will be given by the gradient of the tangent at O. The tangent at O is the line OM and its gradient may be found by using a suitable right-angled triangle at MNO and finding the ratio $\dfrac{MN}{ON}$.

Hence the initial rate of growth of $\quad i = \dfrac{MN}{ON} = \dfrac{2\,\text{amperes}}{0.1\,\text{seconds}}$

$$= 20\,\text{amperes per second}$$

b) The point P on the curve corresponds to a current of 1.0 amperes and the time at which this occurs may be read from the $t$ scale and is 0.07 seconds.

Similarly point Q corresponds to a 1.6 ampere current and occurs at 0.16 seconds.

Hence the time between P and Q = $0.16 - 0.07 = 0.09$ seconds.

This means that the time for the current to increase from 1 to 1.6 amperes is 0.09 seconds.

# Exercise 15.4

1) Plot a graph of $y = e^{2x}$ for values of $x$ from $-1$ to $+1$ at 0.25 unit intervals. Use the graph to find the value of $y$ when $x = 0.3$, and the value of $x$ when $y = 5.4$

2) Using values of $x$ from $-4$ to $+4$ at one unit intervals plot a graph of $y = e^{-x/2}$. Hence find the value of $x$ when $y = 2$, and the gradient of the curve when $x = 0$

3) For a constant pressure process on a certain gas the formula connecting the absolute temperature $T$ and the specific entropy $s$ is $T = 24\,e^{3s}$. Plot a graph of $T$ against $s$ taking values of $s$ equal to 1.000, 1.033, 1.066, 1.100, 1.133, 1.166 and 1.200. Use the graph to find the value of:

(a) $T$ when $s = 1.09$ \qquad (b) $s$ when $T = 700$

4) The equation $i = 2.4e^{-6t}$ gives the relationship between the instantaneous current, $i$ mA, and the time, $t$ seconds. Plot a graph of $i$ against $t$ for values of $t$ from 0 to 0.6 seconds at 0.1 second intervals. Use the curve obtained to find the rate at which the current is decreasing when $t = 0.2$ seconds.

5) In a capacitive circuit the voltage $v$ and the time $t$ seconds are connected by the relationship $v = 240(1 - e^{-5t})$. Draw the curve of $v$ against $t$ for values of $t = 0$ to $t = 0.7$ seconds at 0.1 second intervals. Hence find:

(a) the time when the voltage is 140 volts, and
(b) the initial rate of growth of the voltage when $t = 0$

6) The number of cells, $N$, in a bacterial population in time, $t$ hours, from the commencement of growth is given by $N = 100e^{1.7t}$, find:

(a) the size of the population after 4 hours growth,
(b) the time in which the population increases tenfold from its initial value,
(c) the instantaneous rate of growth after 3 hours from the start.

7) Given that the mass, $m$ grams, of a bacterial population after $t$ hours from the beginning of growth is given by $m = (10^{-10})e^{1.2t}$, find:

(a) the mass of the population after 2 hours from growth commencement,

(b) the mass of the population at beginning of growth,

(c) the time when the population has doubled from its initial value.

8) The decomposition of a chemical compound, $C$, over a period of time, $t$, is given by $C = k(1 - e^{-0.2t})$. If $k = 10$ find the rate of decomposition after 15 seconds.

# GRAPHS OF NON-LINEAR RELATIONSHIPS

After reaching the end of this chapter you should be able to:

1. Draw up suitable tables of values and plot the curves of the types:

$$y = ax^2 + bx + c, \quad y = \frac{a}{x}, \quad y = x^{1/2},$$

$$x^2 + y^2 = a^2, \quad \frac{x^2}{a^2} + \frac{y^2}{b^2} = 1, \quad \frac{x^2}{a^2} - \frac{y^2}{b^2} = 1,$$

$$xy = c^2.$$

2. Recognise the effects on the above curves by changes in the constants a, b, and c.

3. Reduce non-linear physical laws, such as

$$y = ax^2 + b \quad or \quad y = \frac{a}{x} + b, \quad and \quad y = a.x^n$$

(using logarithms), to a straight line graph form.

4. Plot the corresponding straight line graph to verify the law, determine the values of the constants a and b, and find intermediate values.

## GRAPHS OF FAMILIAR EQUATIONS

As technicians it is important that we can recognise the shapes and layout of curves related to their equations. In this section we will draw graphs of the more common equations.

## Graph of $y = ax^2 + bx + c$: Parabola

The important part of the curve is usually the portion in the vicinity of the vertex of the parabola (Fig. 16.1).

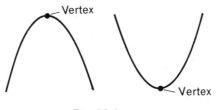

Fig. 16.1

The shape and layout will depend on the values of the constants $a$, $b$ and $c$ and we will examine the effect of each constant in turn.

## Constant *a*

Consider the equation $y = ax^2$. The table of values given below is for $a = 4$, $a = 2$ and $a = -1$:

| $x$ | $-3$ | $-2$ | $-1$ | $0$ | $1$ | $2$ | $3$ |
|---|---|---|---|---|---|---|---|
| $y = 4x^2$ | 36 | 16 | 4 | 0 | 4 | 16 | 36 |
| $y = 2x^2$ | 18 | 8 | 2 | 0 | 2 | 8 | 18 |
| $y = -x^2$ | 9 | 4 | 1 | 0 | 1 | 4 | 9 |

Fig. 16.2 shows the graphs of $y = ax^2$ when $a = 4$, $a = 2$ and $a = -1$

We can see that if the value of $a$ is positive the curve is shaped $\smile$, and the greater the value of $a$, the 'steeper' the curve rises.

Negative values of $a$ give a curve shaped $\frown$.

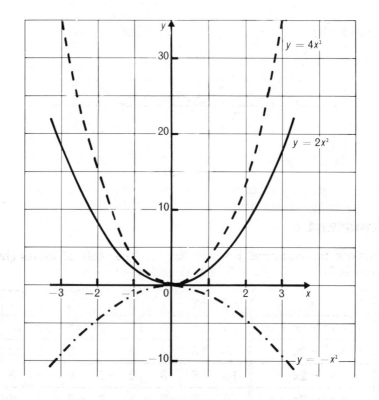

Fig. 16.2

## Constant $b$

Consider the equation $y = x^2 + bx$. The table of values given below is for $b = 2$ and $b = -3$:

| $x$ | $-3$ | $-2$ | $-1$ | 0 | 1 | 2 | 3 | 4 |
|-----|------|------|------|---|---|---|---|---|
| $y = x^2 + 2x$ | 3 | 0 | $-1$ | 0 | 3 | 8 | 15 | 24 |
| $y = x^2 - 3x$ | 18 | 10 | 4 | 0 | $-2$ | $-2$ | 0 | 4 |

Fig. 16.3 shows the graphs of $y = x^2 + bx$ when $b = 2$ and $b = -3$. The effect of a positive value of $b$ is to move the vertex to the left of the vertical $y$-axis, whilst a negative value of $b$ moves the vertex to the right of the vertical axis.

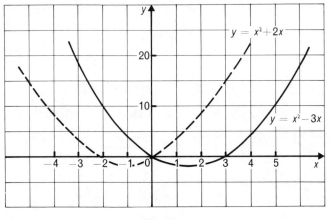

Fig. 16.3

## Constant $c$

Consider the equation $y = x^2 - 2x + c$. The table of values given below is for $c = 10$, $c = 5$, $c = 0$ and $c = -5$

| $x$ | $-3$ | $-2$ | $-1$ | 0 | 1 | 2 | 3 | 4 |
|-----|------|------|------|---|---|---|---|---|
| $y = x^2 - 2x + 10$ | 25 | 18 | 13 | 10 | 9 | 10 | 13 | 18 |
| $y = x^2 - 2x + 5$ | 20 | 13 | 8 | 5 | 4 | 5 | 8 | 13 |
| $y = x^2 - 2x$ | 15 | 8 | 3 | 0 | $-1$ | 0 | 3 | 8 |
| $y = x^2 - 2x - 5$ | 10 | 3 | $-2$ | $-5$ | $-6$ | $-5$ | $-2$ | 3 |

Fig. 16.4 shows the graphs of $y = x^2 - 2x + c$ when $c = 10$, $c = 5$, $c = 0$ and $c = -5$. As we can see the effect is to move the vertex up or down according to the magnitude of $c$.

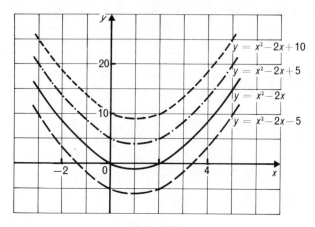

Fig. 16.4

# The graph of $y = ax^{1/2}$    Parabola
## or $y = a\sqrt{x}$: (with horizontal axis)

The table of values given below is for $a = 4$ which is the graph of $y = 4\sqrt{x}$:

| $x$ | All negative values | 0 | 1 | 2 | 3 | 4 | 5 | 6 |
|---|---|---|---|---|---|---|---|---|
| $y = 4\sqrt{x}$ | No real values | 0 | $\pm4$ | $\pm5.66$ | $\pm6.93$ | $\pm8$ | $\pm8.94$ | $\pm9.80$ |

The graph of $y = 4\sqrt{x}$ is shown in Fig. 16.5. We can see that as the value of $a$ is increased so the 'overall depth' of the figure increases.

If the given equation is rearranged we may obtain

$$x = \frac{y^2}{a^2}$$

or $\qquad\qquad x = $ (A constant)$\times y^2$

which is the equation of a parabola with a horizontal axis of symmetry.

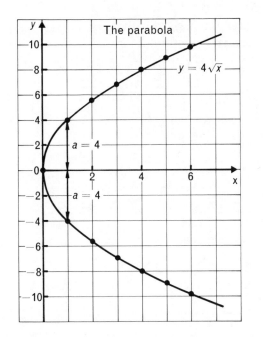

Fig. 16.5

# The graph of $y = \dfrac{a}{x}$: Reciprocal curve (rectangular hyperbola)

The table of values given below is for $a = 4$ which is the graph of $y = \dfrac{4}{x}$:

| $y$ | $-10$ | $-8$ | $-6$ | $-4$ | $-2$ | $-1$ | $-0.5$ | $0$ |
|---|---|---|---|---|---|---|---|---|
| $y = \dfrac{4}{x}$ | $-0.4$ | $-0.5$ | $-0.67$ | $-1$ | $-2$ | $-4$ | $-8$ | |
| | Similar numerical results will occur for positive values | | | | | | | |

The graph of $y = \dfrac{4}{x}$ is shown in Fig. 16.6. We can see that as the value of $a$ decreases the curves are brought nearer to the axes, and vice-versa.

We should also note that for extreme values the curves will become nearer and nearer to the axes, but will never actually 'touch' them. At these extreme values the axes are said to be 'asymptotic' to the curves. In this case the axes are also called 'asymptotes'.

We shall see later on page 268 that the curves comprise, what is called, a 'rectangular hyperbola'.

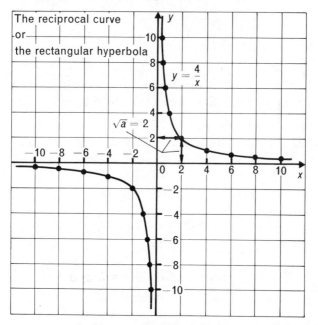

Fig. 16.6

# The graph of $x^2 + y^2 = a^2$: **Circle**

The graph of $x^2 + y^2 = 4$, which is when $a = 2$, is shown in Fig. 16.7.

The table of values has been omitted here but the reader may find it useful to rearrange the equation and plot the curve.

We can see that the radius of the circle is given by the value of constant $a$.

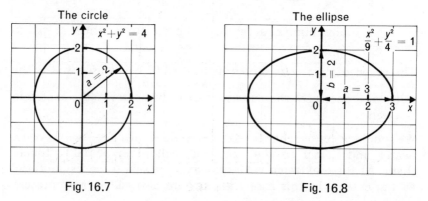

Fig. 16.7                                           Fig. 16.8

# The graph of $\dfrac{x^2}{a^2}+\dfrac{y^2}{b^2}=1$:   Ellipse

The graph of $\dfrac{x^2}{9}+\dfrac{y^2}{4}=1$, which is when $a=3$ and $b=2$, is shown in Fig. 16.8.

The line along which the greatest dimension (given by $2a$) across the ellipse is measured is called the major axis. In this case it lies along the $x$-axis.

The line along which the least dimension (given by $2b$) across the ellipse is measured is called the minor axis. In this case it lies along the $y$-axis.

# The graph of $\dfrac{x^2}{a^2}-\dfrac{y^2}{b^2}=1$:   Hyperbola

The graph of $\dfrac{x^2}{9}-\dfrac{y^2}{4}=1$, which is when $a=3$ and $b=2$, is shown in Fig. 16.9.

The ratio $\pm\dfrac{b}{a}$, in this case $\pm\dfrac{2}{3}$, gives the slopes of two straight lines called asymptotes. At extreme values these lines will become nearer and nearer to the curves but will never actually 'touch' them, and are said to be *asymptotic* to the curves.

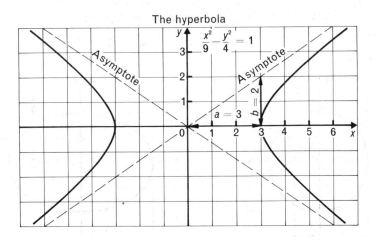

The hyperbola

Fig. 16.9

# The graph of $xy = c^2$: Rectangular hyperbola

The graph of $xy = 9$, which is when $c = 3$, is shown in Fig. 16.10.

We can see that the curve is similar to the reciprocal curve discussed earlier. It is also similar to the ordinary hyperbola, the description 'rectangular' referring to the fact that the asymptotes are also the rectangular axes.

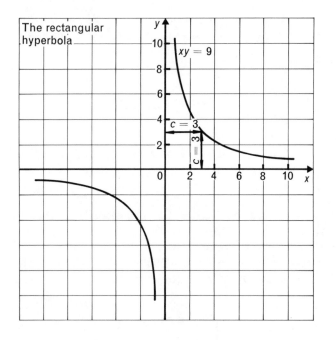

Fig. 16.10

# Exercise 16.1

State which answer or answers are correct in Questions 1–9. In every diagram the origin is at the intersection of the axes.

1) The graph of $y = x^2$ is:

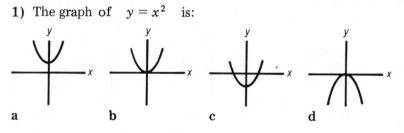

a　　　　　b　　　　　c　　　　　d

**2)** The graph of $y = x^2 + 2$ is:

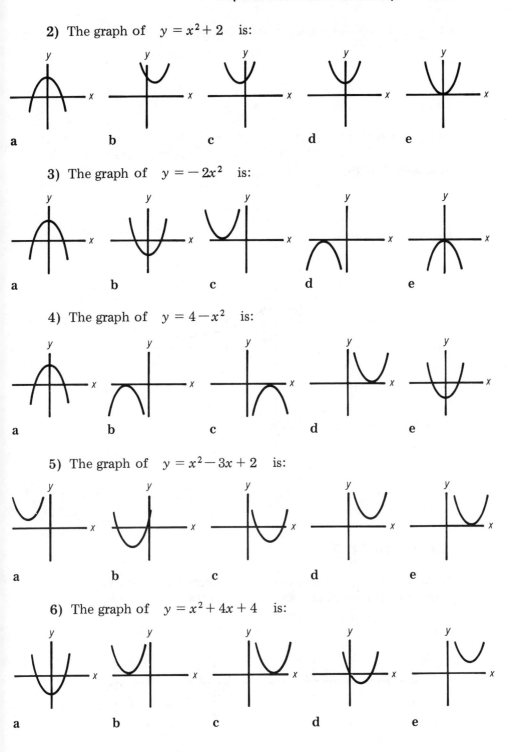

a          b          c          d          e

**3)** The graph of $y = -2x^2$ is:

a          b          c          d          e

**4)** The graph of $y = 4 - x^2$ is:

a          b          c          d          e

**5)** The graph of $y = x^2 - 3x + 2$ is:

a          b          c          d          e

**6)** The graph of $y = x^2 + 4x + 4$ is:

a          b          c          d          e

7) The graph of $\dfrac{x^2}{16} + \dfrac{y^2}{9} = 1$   is:

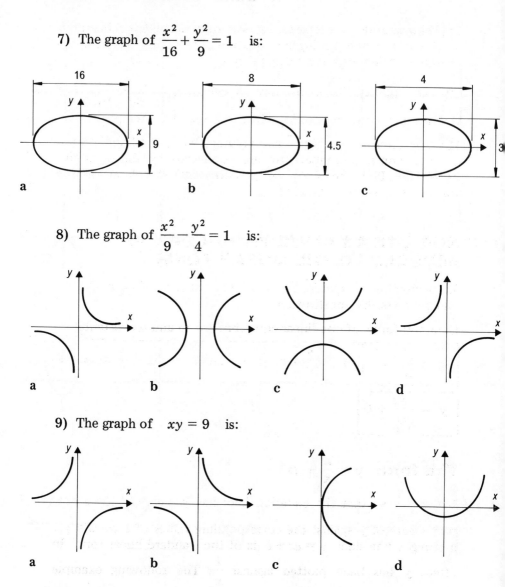

a

b

c

8) The graph of $\dfrac{x^2}{9} - \dfrac{y^2}{4} = 1$   is:

a

b

c

d

9) The graph of   $xy = 9$   is:

a

b

c

d

10) What is the equation of the circle shown in the diagram?

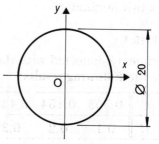

**11)** The profile of the cross-section of a headlamp reflector is parabolic in shape, having an equation $y = 0.00617x^2$. Plot the profile for values of $x$ from 0 to 90 mm.

**12)** An elliptical template has an equation: $\dfrac{x^2}{3600} + \dfrac{y^2}{1600} = 1$ where $x$ and $y$ have mm units. Plot the shape of the template.

**13)** The pressure, $p$ MN/m$^2$, and volume, $V$m$^3$, of a fixed mass of air at a constant temperature are connected by the equation: $pV = 0.16$. Plot the graph of pressure against volume, as the volume changes from 0 to 1 m$^3$.

## NON-LINEAR LAWS WHICH CAN BE REDUCED TO THE LINEAR FORM

Many non-linear equations can be reduced to the linear form by making a suitable substitution.

Common forms of non-linear equations are ($a$ and $b$ constants):

| | | | |
|---|---|---|---|
| $y = \dfrac{a}{x} + b$ | $y = \dfrac{a}{x^2} + b$ | $y = ax^2 + b$ | $y = a\sqrt{x} + b$ |

| |
|---|
| $y = \dfrac{a}{\sqrt{x}} + b$ |

## The form $y = \dfrac{a}{x} + b$

Let $z = \dfrac{1}{x}$ so that the equation becomes $y = az + b$. If we now plot values of $y$ against the corresponding values of $z$ we will get a straight line since $y = az + b$ is of the standard linear form. In effect $y$ has been plotted against $\dfrac{1}{x}$. The following example illustrates this method.

### EXAMPLE 16.1

An experiment connected with the flow of water over a rectangular weir gave the following results:

| $C$ | 0.503 | 0.454 | 0.438 | 0.430 | 0.425 | 0.421 |
|---|---|---|---|---|---|---|
| $H$ | 0.1 | 0.2 | 0.3 | 0.4 | 0.5 | 0.6 |

The relation between $C$ and $H$ is thought to be of the form $C = \dfrac{a}{H} + b$. Test if this is so and find the values of the constants $a$ and $b$.

In the suggested equation $C$ is the sum of two terms, the first of which varies as $\dfrac{1}{H}$. If the equation $C = \dfrac{a}{H} + b$ is correct then when we plot $C$ against $\dfrac{1}{H}$ we should obtain a straight line. To do this we draw up the following table:

| $C$ | 0.503 | 0.454 | 0.438 | 0.430 | 0.425 | 0.421 |
|---|---|---|---|---|---|---|
| $\dfrac{1}{H}$ | 10.00 | 5.00 | 3.33 | 2.50 | 2.00 | 1.67 |

The graph obtained is shown in Fig. 16.11. It is a straight line and hence the given values follow a law of the form $C = \dfrac{a}{H} + b$.

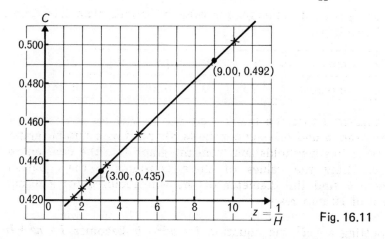

Fig. 16.11

To find the values of $a$ and $b$ we choose two points which lie on the straight line.

The point (3.00, 0.435) lies on the line.

$\therefore$ $\qquad\qquad 0.435 = 3.00a + b$ $\qquad\qquad$ [1]

The point (9.00, 0.492) also lies on the line.

$\therefore$ $\qquad\qquad 0.492 = 9.00a + b$ $\qquad\qquad$ [2]

Subtracting equation [1] from equation [2] gives

$$0.492 - 0.435 = a(9.00 - 3.00)$$

$$\therefore \qquad a = 0.0095$$

Substituting this value for $a$ in equation [1] gives

$$0.435 = 3.00 \times 0.0095 + b$$

$$\therefore \qquad b = 0.435 - 0.0285 = 0.407$$

Hence the values of $a$ and $b$ are 0.0095 and 0.407 respectively.


# The form $y = ax^2 + b$

Let $z = x^2$ and as previously if we plot values of $y$ against $z$ (in effect $x^2$) we will get a straight line since $y = az + b$ is of the standard form. The following example illustrates this method.


### EXAMPLE 16.2

The fusing current $I$ amperes for wires of various diameters $d$ mm is as shown below:

| $d$ (mm) | 5 | 10 | 15 | 20 | 25 |
|---|---|---|---|---|---|
| $I$ (amperes) | 6.25 | 10 | 16.25 | 25 | 36.25 |

It is suggested that the law $I = ad^2 + b$ is true for the range of values given, $a$ and $b$ being constants. By plotting a suitable graph show that this law holds and from the graph find the constants $a$ and $b$. Using the values of these constants in the equation $I = ad^2 + b$ find the diameter of the wire required for a fusing current of 12 amperes.

By putting $z = d^2$ the equation $I = ad^2 + b$ becomes $I = az + b$ which is the standard form of a straight line. Hence by plotting $I$ against $d^2$ we should get a straight line if the law is true. To try this we draw up a table showing corresponding values of $I$ and $d^2$.

| $d$ | 5 | 10 | 15 | 20 | 25 |
|---|---|---|---|---|---|
| $z = d^2$ | 25 | 100 | 225 | 400 | 625 |
| $I$ | 6.25 | 10 | 16.25 | 25 | 36.25 |

From the graph (Fig. 16.12) we see that the points do lie on a straight line and hence the values obey a law of the form $I = ad^2 + b$.

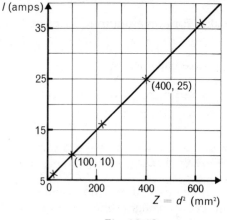

Fig. 16.12

To find the values of $a$ and $b$ choose two points which lie on the line and find their co-ordinates.

The point $(400, 25)$ lies on the line

$$\therefore \qquad 25 \;=\; 400a + b \qquad\qquad [1]$$

The point $(100, 10)$ lies on the line

$$\therefore \qquad 10 \;=\; 100a + b \qquad\qquad [2]$$

Subtracting equation [2] from equation [1] gives

$$15 \;=\; 300a$$

$$\therefore \qquad a \;=\; 0.05$$

Substituting $a = 0.05$ in equation [2] gives

$$10 \;=\; 100 \times 0.05 + b$$

$$\therefore \qquad b \;=\; 5$$

Therefore the law is

$$I \;=\; 0.05d^2 + 5$$

When $I = 12$

$$12 \;=\; 0.05d^2 + 5$$

$$\therefore \qquad d \;=\; \sqrt{140} \;=\; 11.8\,\text{mm}$$

# Consider $y = \dfrac{a}{x^2} + b$

Let $z = \dfrac{1}{x^2}$ so that the equation becomes $y = az + b$. If we now plot values of $y$ against corresponding values of $z$ we will get a straight line since $y = az + b$ is of the standard linear form. In effect $y$ has been plotted against $\dfrac{1}{x^2}$.

# Consider $y = a\sqrt{x} + b$

Let $z = \sqrt{x}$ and as previously, if we plot values of $y$ against $z$ (in effect $\sqrt{x}$) we will obtain a straight line since $y = az + b$ is of the standard linear form.

# The form $y = \dfrac{a}{\sqrt{x}} + b$

This may also be written equivalently as $y = \dfrac{a}{x^{1/2}} + b$ or $y = ax^{-1/2} + b$.

Let $z = \dfrac{1}{\sqrt{x}}$ and as previously, if we plot values of $y$ against $z$ $\left(\text{in effect } \dfrac{1}{\sqrt{x}}\right)$ we will obtain a straight line since $y = az + b$ is the standard linear form.

# Exercise 16.2

1) The following readings were taken during a test:

| $R$ (ohms) | 85 | 73.3 | 64 | 58.8 | 55.8 |
|---|---|---|---|---|---|
| $I$ (amperes) | 2 | 3 | 5 | 8 | 12 |

$R$ and $I$ are thought to be connected by an equation of the form $R = \dfrac{a}{I} + b$.

Verify that this is so by plotting $R$ ($y$-axis) against $\dfrac{1}{I}$ ($x$-axis) and hence find values for $a$ and $b$.

**2)** In the theory of the moisture content of thermal insulation efficiency of porous materials the following table gives values of $\mu$, the diffusion constant of the material, and $k_m$, the thermal conductivity of damp insulation material:

| $\mu$ | 1.3 | 2.7 | 3.8 | 5.4 | 7.2 | 10.0 |
|-------|------|------|------|------|------|------|
| $k_m$ | 0.0336 | 0.0245 | 0.0221 | 0.0203 | 0.0192 | 0.0183 |

Find the equation connecting $\mu$ and $k_m$ if it is of the form $k_m = a + \dfrac{b}{\mu}$ where $a$ and $b$ are constants.

**3)** The accompanying table gives the corresponding values of the pressure, $p$, of mercury and the volume, $v$, of a given mass of gas at constant temperature.

| $p$ | 90 | 100 | 130 | 150 | 170 | 190 |
|-----|------|-------|-------|------|------|------|
| $v$ | 16.66 | 13.64 | 11.54 | 9.95 | 8.82 | 7.89 |

By plotting $p$ against the reciprocal of $v$ obtain some relation between $p$ and $v$. Evaluate any constants used in your method.

**4)** The approximate number of a type of bacteria, $B$, is checked regularly and recorded in the table below.

| Bacteria, $B (\times 10^3)$ | 5 | 28.5 | 41.0 | 113.0 | 253.6 | 450.0 |
|-----------------------------|-----|------|------|-------|-------|-------|
| Time, $t$ (hours) | 1.0 | 2.5 | 3.0 | 5.0 | 7.5 | 10.0 |

It is thought that the growth is related according to the law $B = mt^2 + c$ where $m$ and $c$ are constants. By plotting a suitable graph verify this to be true and evaluate $m$ and $c$.

**5)** In an experiment, the resistance $R$ of copper wire of various diameters $d$ mm was measured and the following readings were obtained.

| $d$ (mm) | 0.1 | 0.2 | 0.3 | 0.4 | 0.5 |
|----------|-----|-----|-----|-----|-----|
| $R$ (ohms) | 20 | 5 | 2.2 | 1.3 | 0.8 |

Show that $R = \dfrac{k}{d^2}$ and find a suitable value for $k$.

6) The following table gives the thickness $T$ mm of a brass flange brazed to a copper pipe of internal diameter $D$ mm:

| $T$ mm | 15.5 | 17.8 | 19.5 | 20.9 | 22.2 | 23.3 |
|--------|------|------|------|------|------|------|
| $D$ mm | 50 | 100 | 150 | 200 | 250 | 300 |

Show that $T$ and $D$ are connected by an equation of the form $T = a\sqrt{D} + b$, find the values of constants $a$ and $b$, and find the thickness of the flange for a 70 mm diameter pipe.

7) The table shows how the coefficient of friction, $\mu$, between a belt and a pulley varies with the speed, $v$ m/s, of the belt. By plotting a graph show that $\mu = m\sqrt{v} + c$ and find the values of constants $m$ and $c$.

| $\mu$ | 0.26 | 0.29 | 0.32 | 0.35 | 0.38 |
|-------|------|------|------|------|------|
| $v$ | 2.22 | 5.00 | 8.89 | 13.89 | 20.00 |

8) Using the table below show that the values are in agreement with the law $y = \dfrac{m}{\sqrt{x}} + c$. Hence evaluate the constants $m$ and $c$.

| $x$ | 0.2 | 0.8 | 1.2 | 1.8 | 2.5 | 4.4 |
|-----|-----|-----|-----|-----|-----|-----|
| $y$ | 1.62 | 1.51 | 1.49 | 1.47 | 1.46 | 1.44 |

9) On a test bed a projectile, starting from rest over a constant measured distance, is subjected to different accelerations. The table below records these accelerations, $A$, and the times, $t$, taken to complete the distance.

| $A$ (ms$^{-2}$) | 5 | 10 | 15 | 20 | 25 | 30 |
|-----------------|---|----|----|----|----|----|
| $t$ (s) | 5.00 | 3.37 | 2.65 | 2.20 | 1.90 | 1.70 |

If the law connecting $A$ and $t$ is thought to be of the form $t = aA^{-1/2} + b$ plot a suitable straight line graph, and find the constants $a$ and $b$.

# REDUCING TO THE LOGARITHMIC FORM

Consider the following relationship, in which $z$ and $t$ are the variables whilst $a$ and $n$ are constants.

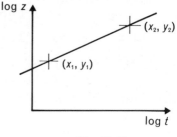

$$z = a.t^n$$

Taking logs we get:

$$\log z = \log (t^n .a)$$
$$= \log t^n + \log a$$
$$\therefore \ \log z = n.(\log t) + \log a$$

Fig. 16.13

Comparing this with the equation $y = mx + c$ of a straight line, we see that if we plot $\log z$ on the $y$-axis and $\log t$ on the $x$-axis (Fig. 16.13) then the result will be a straight line.

Two convenient points, $(x_1, y_1)$ and $(x_2, y_2)$, which lie on the line should be chosen. The co-ordinates are now substituted in the straight line equation:

The point $(x_1, y_1)$ lies on the line — thus    $y_1 = nx_1 + \log a$

and point $(x_2, y_2)$ lies on the line — thus    $y_2 = nx_2 + \log a$

These two equations may be solved simultaneously to find the values of $n$ and $a$.

### EXAMPLE 16.3

The law connecting two quantities $z$ and $t$ is of the form $z = a.t^n$ Find the values of the constants $a$ and $n$ given the following pairs of values:

| $z$ | 3.170 | 4.603 | 7.499 | 10.50 | 15.17 |
|---|---|---|---|---|---|
| $t$ | 7.980 | 9.863 | 13.03 | 15.81 | 19.50 |

By taking logs and rearranging (see text) we have

$$\log z = n.\log t + \log a \qquad\qquad [1]$$

For the numerical part of the solution we may use common logarithms (logs to the base 10) or natural logarithms (logs to the base e).

The solution given uses common logarithms. The reader may find it instructive to work through this example using natural logarithms and verify that the same results are obtained.

From the given values, using logarithms to the base 10:

| $\log_{10}z$ | 0.5011 | 0.6631 | 0.8750 | 1.0212 | 1.1810 |
|---|---|---|---|---|---|
| $\log_{10}t$ | 0.9020 | 0.9940 | 1.1149 | 1.1990 | 1.2900 |

Since it is not convenient to show the origin (point $0, 0$) we shall use the two-point method of finding the constants (Fig. 16.14).

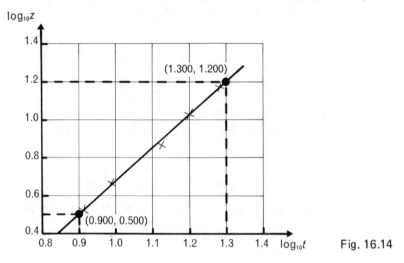

Fig. 16.14

Point $(0.900, 0.500)$ lies on the line, and substituting in equation [1] gives

$$0.500 = n(0.900) + \log_{10}a \qquad [2]$$

Point $(1.300, 1.200)$ lies on the line, and substituting in equation [1] gives

$$1.200 = n(1.300) + \log_{10}a \qquad [3]$$

Subtracting equation [2] from equation [3],

$$0.7 = 0.4n \qquad \therefore \; n = 1.75$$

Substituting in the equation [2] gives

$$0.500 = 1.75(0.900) + \log_{10}a$$

$$\therefore \qquad \log_{10}a = -1.075$$

$$\therefore \qquad a = 0.084$$

## Exercise 16.3

**1)** The following values of $x$ and $y$ follow a law of the type $y = ax^n$. By plotting $\log y$ (vertically) against $\log x$ (horizontally) find values for $a$ and $n$.

| $x$ | 1 | 2 | 3 | 4 | 5 |
|---|---|---|---|---|---|
| $y$ | 3 | 12 | 27 | 48 | 75 |

**2)** The following results were obtained in an experiment to find the relationships between the luminosity $I$ of a metal filament lamp and the voltage $V$.

| $V$ | 40 | 60 | 80 | 100 | 120 |
|---|---|---|---|---|---|
| $I$ | 5.1 | 26.0 | 82 | 200 | 414 |

The law is thought to be of the type $I = aV^n$ Test this by plotting $\log I$ (vertically) against $\log V$ (horizontally) and find suitable values for $a$ and $n$.

**3)** The relationship between power $P$ (watts), the e.m.f. $E$ (volts) and the resistance $R$ (ohms) is thought to be of the form $P = \dfrac{E^n}{R}$. In an experiment in which $R$ was kept constant the following results were obtained:

| $E$ | 5 | 10 | 15 | 20 | 25 | 30 |
|---|---|---|---|---|---|---|
| $P$ | 2.5 | 10 | 22.5 | 40 | 62.5 | 90 |

Verify the law and find the values of the constants $n$ and $R$.

**4)** The following results were obtained in an experiment to find the relationship between the luminosity $I$ of a metal filament lamp and the voltage $V$.

| $V$ | 60 | 80 | 100 | 120 | 140 |
|---|---|---|---|---|---|
| $I$ | 11 | 20.5 | 89 | 186 | 319 |

Allowing for the fact that an error was made in one of the readings show that the law between $I$ and $V$ is of the form $I = aV^n$ and find the probable correct value of the reading. Find the value of $n$.

**5)** Two quantities $t$ and $m$ are connected by a law of the type $t = a.m^b$ and the co-ordinates of two points which satisfy the equation are $(8, 6.8)$ and $(20, 26.9)$. Find the law.

**6)** The intensity of radiation, $R$, from certain radioactive materials at a particular time $t$ is thought to follow the law $R = kt^n$. In an experiment to test this the following values were obtained:

| $R$ | 58 | 43.5 | 26.5 | 14.5 | 10 |
|---|---|---|---|---|---|
| $t$ | 1.5 | 2 | 3 | 5 | 7 |

Show that the assumption was correct and evaluate $k$ and $n$.

# 17.

# MATRICES

After reaching the end of this chapter you should be able to:

1. Recognise the notation of a matrix.
2. Calculate the sum and difference of two matrices (2 × 2 only).
3. Calculate the product of two 2 × 2 matrices.
4. Demonstrate that the product of two matrices is, in general, non-commutative.
5. Define the unit matrix.
6. Recognise the notation for a determinant.
7. Evaluate a 2 × 2 determinant.
8. Solve simultaneous linear equations with two unknowns using determinants.
9. Describe the meaning of a determinant whose value is zero and define a singular matrix.
10. Obtain the inverse of a 2 × 2 matrix.
11. Solve simultaneous linear equations with two unknowns by means of matrices.

## INTRODUCTION

The block within which a printer sets his type, and a car radiator could each be called a matrix.

A matrix in mathematics is any rectangular array of numbers, usually enclosed in brackets.

Before using matrices in practical problems we first look at how they are ordered, added, subtracted and manipulated in other ways.

## ELEMENT

Each number or symbol in a matrix is called an *element* of the matrix.

## ORDER

The *dimension* or *order* of a matrix is stated by the number of rows followed by the number of columns in the rectangular array.

e.g.

| Matrix | $\begin{pmatrix} 1 & 2 \\ 3 & 4 \end{pmatrix}$ | $\begin{pmatrix} a & 2 & -3 \\ 4 & b & x \end{pmatrix}$ | $\begin{pmatrix} \sin \theta & 1 \\ \cos \theta & 2 \\ \tan \theta & 3 \end{pmatrix}$ | $(6)$ |
|---|---|---|---|---|
| Order | $2 \times 2$ | $2 \times 3$ | $3 \times 2$ | $1 \times 1$ |

# EQUALITY

If two matrices are equal, then they must be of the same order and their corresponding elements must be equal.

Thus if $\begin{pmatrix} 2 & 3 & x \\ a & 5 & -2 \end{pmatrix} = \begin{pmatrix} 2 & 3 & 4 \\ -1 & 5 & -2 \end{pmatrix}$ then $x = 4$ and $a = -1$

# ADDITION AND SUBTRACTION

Two matrices may be added or subtracted only if they are of the *same order*. We say the matrices are *conformable* for addition (or subtraction) and we add (or subtract) by combining corresponding elements.

### EXAMPLE 17.1

If $A = \begin{pmatrix} 3 & 4 \\ 5 & 6 \end{pmatrix}$ and $B = \begin{pmatrix} 0 & 6 \\ 5 & 2 \end{pmatrix}$ determine:  a) $C = A + B$
and  b) $D = A - B$

a)  $C = \begin{pmatrix} 3 & 4 \\ 5 & 6 \end{pmatrix} + \begin{pmatrix} 0 & 6 \\ 5 & 2 \end{pmatrix} = \begin{pmatrix} 3+0 & 4+6 \\ 5+5 & 6+2 \end{pmatrix} = \begin{pmatrix} 3 & 10 \\ 10 & 8 \end{pmatrix}$

b)  $D = \begin{pmatrix} 3 & 4 \\ 5 & 6 \end{pmatrix} - \begin{pmatrix} 0 & 6 \\ 5 & 2 \end{pmatrix} = \begin{pmatrix} 3-0 & 4-6 \\ 5-5 & 6-2 \end{pmatrix} = \begin{pmatrix} 3 & -2 \\ 0 & 4 \end{pmatrix}$

# ZERO OR NULL MATRIX

A *zero* or *null* matrix, denoted by $O$, is one in which all the elements are zero. It may be of any order.

Thus $\begin{pmatrix} 0 & 0 \\ 0 & 0 \end{pmatrix}$ is a zero matrix of order 2. It behaves like zero in the real number system.

# IDENTITY OR UNIT MATRIX

The *identity* matrix can be of any suitable order with all the main diagonal elements 1 and the remaining elements 0. It is denoted by $I$ and behaves like unity in the real number system.

Thus $\begin{pmatrix} 1 & 0 \\ 0 & 1 \end{pmatrix}$ is a unit matrix of order 2

# TRANSPOSE

The *transpose* of a matrix $A$ is written as $A'$ or $A^T$. When the row of a matrix is interchanged with its corresponding column, that is row 1 becomes column 1 and row 2 becomes column 2 and so on then the matrix is transposed.

Thus if $A = \begin{pmatrix} 1 & 2 & -3 \\ 4 & 7 & 0 \end{pmatrix}$ then $A' = \begin{pmatrix} 1 & 4 \\ 2 & 7 \\ -3 & 0 \end{pmatrix}$

## Exercise 17.1

**1)** State the order of each of the following matrices:

(a) $\begin{pmatrix} 1 & 2 \\ 3 & 4 \end{pmatrix}$
(b) $\begin{pmatrix} 5 \\ -6 \end{pmatrix}$
(c) $\begin{pmatrix} a & b & 4 \\ 2 & 3 & 5 \\ x & -6 & 0 \end{pmatrix}$

(d) $\begin{pmatrix} 1 & -2 & -3 & -4 \\ 6 & 2 & 0 & -1 \end{pmatrix}$

**2)** How many elements are there in:

(a) a $3 \times 3$ matrix          (b) a $2 \times 2$ matrix
(c) a square matrix of order $n$?

**3)** Write down the transpose of each matrix in Question 1).

**4)** Combine the following matrices:

(a) $\begin{pmatrix} 2 & 1 \\ 3 & 2 \end{pmatrix} + \begin{pmatrix} -2 & -1 \\ 6 & 0 \end{pmatrix}$
(b) $\begin{pmatrix} 2 & 1 \\ 3 & 2 \end{pmatrix} - \begin{pmatrix} -2 & -1 \\ 6 & 0 \end{pmatrix}$

(c) $\begin{pmatrix} \frac{1}{2} & 1 \\ \frac{1}{3} & \frac{1}{5} \end{pmatrix} + \begin{pmatrix} \frac{1}{3} & -\frac{1}{2} \\ \frac{1}{2} & \frac{4}{5} \end{pmatrix}$

**5)** Determine $a, b$ and $c$ if $(a \quad b \quad c) - (-3 \quad 4 \quad 1) = (-5 \quad 1 \quad 0)$.

**6)** Complete $\begin{pmatrix} \frac{1}{2} & \frac{1}{4} \\ \frac{1}{5} & \frac{1}{6} \end{pmatrix} - \begin{pmatrix} \frac{1}{6} & \frac{1}{5} \\ \frac{1}{6} & \frac{1}{9} \end{pmatrix}$

**7)** Solve the equation $X - \begin{pmatrix} 1 & 3 \\ 5 & -2 \end{pmatrix} = \begin{pmatrix} 4 & 5 \\ 7 & 0 \end{pmatrix}$ where $X$ is a $2 \times 2$ matrix.

**8)** If $\begin{pmatrix} 4 \\ 5 \end{pmatrix} + \begin{pmatrix} x \\ y \end{pmatrix} = \begin{pmatrix} 4 \\ 10 \end{pmatrix}$, determine $\begin{pmatrix} x \\ y \end{pmatrix}$

# MULTIPLICATION OF A MATRIX BY A REAL NUMBER

A matrix may be multiplied by a number in the following way:

$$4 \begin{pmatrix} 2 & 3 \\ 7 & -1 \end{pmatrix} = \begin{pmatrix} 4 \times 2 & 4 \times 3 \\ 4 \times 7 & 4 \times (-1) \end{pmatrix} = \begin{pmatrix} 8 & 12 \\ 28 & -4 \end{pmatrix}$$

Conversely the common factor of each element in a matrix may be written outside the matrix. Thus $\begin{pmatrix} 9 & 3 \\ 42 & 15 \end{pmatrix} = 3 \begin{pmatrix} 3 & 1 \\ 14 & 5 \end{pmatrix}$

# MATRIX MULTIPLICATION

Two matrices can only be multiplied together if the number of columns in the first matrix is equal to the number of rows in the second matrix. We say that the matrices are *conformable* for multiplication. The method for multiplying together a pair of $2 \times 2$ matrices is as follows

$$\begin{pmatrix} a & b \\ c & d \end{pmatrix} \times \begin{pmatrix} e & f \\ g & h \end{pmatrix} = \begin{pmatrix} ae + bg & af + bh \\ ce + dg & cf + dh \end{pmatrix}$$

**EXAMPLE 17.2**

a) $\begin{pmatrix} 2 & 3 \\ 4 & 5 \end{pmatrix} \times \begin{pmatrix} 7 & 1 \\ 0 & 6 \end{pmatrix} = \begin{pmatrix} (2 \times 7) + (3 \times 0) & (2 \times 1) + (3 \times 6) \\ (4 \times 7) + (5 \times 0) & (4 \times 1) + (5 \times 6) \end{pmatrix}$

$$= \begin{pmatrix} 14 & 20 \\ 28 & 34 \end{pmatrix}$$

b) $\begin{pmatrix} 3 & 4 \\ 2 & 5 \end{pmatrix} \times \begin{pmatrix} 6 \\ 7 \end{pmatrix} = \begin{pmatrix} (3 \times 6) + (4 \times 7) \\ (2 \times 6) + (5 \times 7) \end{pmatrix} = \begin{pmatrix} 46 \\ 47 \end{pmatrix}$

c) $\begin{pmatrix} 3 \\ 2 \end{pmatrix} \times \begin{pmatrix} 4 & 6 \\ 5 & 7 \end{pmatrix}$    This is not possible since the matrices are not comformable.

**EXAMPLE 17.3**

Form the products $AB$ and $BA$ given that $A = \begin{pmatrix} 1 & 2 \\ 3 & 4 \end{pmatrix}$ and

$B = \begin{pmatrix} 5 & 6 \\ 7 & 8 \end{pmatrix}$ and hence show that $AB \neq BA$.

$$AB = \begin{pmatrix} 1 & 2 \\ 3 & 4 \end{pmatrix}\begin{pmatrix} 5 & 6 \\ 7 & 8 \end{pmatrix} = \begin{pmatrix} (1 \times 5)+(2 \times 7) & (1 \times 6)+(2 \times 8) \\ (3 \times 5)+(4 \times 7) & (3 \times 6)+(4 \times 8) \end{pmatrix}$$

$$= \begin{pmatrix} 19 & 22 \\ 43 & 50 \end{pmatrix}$$

$$BA = \begin{pmatrix} 5 & 6 \\ 7 & 8 \end{pmatrix}\begin{pmatrix} 1 & 2 \\ 3 & 4 \end{pmatrix} = \begin{pmatrix} (5 \times 1)+(6 \times 3) & (5 \times 2)+(6 \times 4) \\ (7 \times 1)+(8 \times 3) & (7 \times 2)+(8 \times 4) \end{pmatrix}$$

$$= \begin{pmatrix} 23 & 34 \\ 31 & 46 \end{pmatrix}$$

As we see the results are different and, in general, matrix multiplication is non-commutative.

## Exercise 17.2

1) If $A = \begin{pmatrix} 3 & 0 \\ -2 & 1 \end{pmatrix}$ and $B = \begin{pmatrix} -4 & 1 \\ 3 & -2 \end{pmatrix}$ determine:

(a) $2A$          (b) $3B$          (c) $2A+3B$          (d) $2A-3B$

2) Calculate the following products:

(a) $\begin{pmatrix} 3 & 1 \\ 2 & 0 \end{pmatrix}\begin{pmatrix} 4 & -1 \\ 2 & 3 \end{pmatrix}$          (b) $\begin{pmatrix} 2 & 1 \\ 3 & 1 \end{pmatrix}\begin{pmatrix} 1 & 0 \\ 0 & 1 \end{pmatrix}$          (c) $\begin{pmatrix} 2 & 1 \\ 4 & 2 \end{pmatrix}\begin{pmatrix} 2 & 3 \\ 1 & 5 \end{pmatrix}$

(d) $\begin{pmatrix} 1 & 0 \\ 0 & 1 \end{pmatrix}\begin{pmatrix} a & b \\ c & d \end{pmatrix}$          (e) $\begin{pmatrix} k & 0 \\ 0 & k \end{pmatrix}\begin{pmatrix} a & b \\ c & d \end{pmatrix}$

3) If $A = \begin{pmatrix} 1 & 2 \\ 3 & 4 \end{pmatrix}$ and $B = \begin{pmatrix} 2 & -1 \\ 1 & 3 \end{pmatrix}$ calculate:

(a) $A^2$ (that is $A \times A$)          (b) $B^2$          (c) $2AB$

(d) $A^2+B^2+2AB$          (e) $(A+B)^2$

## DETERMINANT OF A SQUARE MATRIX OF ORDER 2

If matrix $A = \begin{pmatrix} a & b \\ c & d \end{pmatrix}$ then its *determinant* is denoted by $|A|$ or $\det A$ and the result is a *number* given by

$$|A| = \begin{vmatrix} a & b \\ c & d \end{vmatrix} = ad - bc$$

**EXAMPLE 17.4**

Evaluate $|A|$ if $A = \begin{pmatrix} 1 & -2 \\ 3 & 4 \end{pmatrix}$

$$|A| = \begin{vmatrix} 1 & -2 \\ 3 & 4 \end{vmatrix} = 1 \times 4 - (-2) \times 3 = 10$$

# SOLUTION OF SIMULTANEOUS LINEAR EQUATIONS USING DETERMINANTS

To solve simultaneous linear equations with two unknowns using determinants, the following procedure is used.

**(1)** Write out the two equations in order:    $a_1x + b_1y = c_1$

$$a_2x + b_2y = c_2$$

**(2)** Calculate $\Delta = \begin{vmatrix} a_1 & b_1 \\ a_2 & b_2 \end{vmatrix}$

**(3)** Then $x = \dfrac{\begin{vmatrix} c_1 & b_1 \\ c_2 & b_2 \end{vmatrix}}{\Delta}$   and   $y = \dfrac{\begin{vmatrix} a_1 & c_1 \\ a_2 & c_2 \end{vmatrix}}{\Delta}$

**EXAMPLE 17.5**

By using determinants, solve the simultaneous equations

$$3x + 4y = 22$$
$$2x + 5y = 24$$

Now    $\Delta = \begin{vmatrix} 3 & 4 \\ 2 & 5 \end{vmatrix} = (3 \times 5) - (4 \times 2) = 7$

Thus    $x = \dfrac{\begin{vmatrix} 22 & 4 \\ 24 & 5 \end{vmatrix}}{7} = \dfrac{(22 \times 5) - (4 \times 24)}{7} = \dfrac{14}{7} = 2$

And    $y = \dfrac{\begin{vmatrix} 3 & 22 \\ 2 & 24 \end{vmatrix}}{7} = \dfrac{(3 \times 24) - (22 \times 2)}{7} = \dfrac{28}{7} = 4$

## Exercise 17.3

1) Evaluate the following determinants:

(a) $\begin{vmatrix} 5 & 2 \\ 3 & 6 \end{vmatrix}$    (b) $\begin{vmatrix} 7 & 4 \\ 5 & 2 \end{vmatrix}$    (c) $\begin{vmatrix} 6 & 8 \\ 2 & 5 \end{vmatrix}$

2) Solve the following simultaneous equations by using determinants:

(a) $3x + 4y = 11$
    $x + 7y = 15$

(b) $5x + 3y = 29$
    $4x + 7y = 37$

(c) $4x - 6y = -2.5$
    $7x - 5y = -0.25$

# THE INVERSE OF A SQUARE MATRIX OF ORDER 2

Instead of dividing a number by 5 we can multiply by $\frac{1}{5}$ and obtain the same result.

Thus $\frac{1}{5}$ is the multiplicative inverse of 5. That is $5 \times \frac{1}{5} = 1$

In matrix algebra we never divide by a matrix but multiply instead by the inverse. The inverse of matrix $A$ is denoted by $A^{-1}$ and is such that

$$AA^{-1} = \begin{pmatrix} 1 & 0 \\ 0 & 1 \end{pmatrix} = I, \quad \text{the identity matrix.}$$

To find the inverse, $A^{-1}$, of the square matrix $A = \begin{pmatrix} a & b \\ c & d \end{pmatrix}$

we use the expression: $A^{-1} = \dfrac{1}{|A|} \begin{pmatrix} d & -b \\ -c & a \end{pmatrix} = \dfrac{1}{ad - bc} \begin{pmatrix} d & -b \\ -c & a \end{pmatrix}$

**EXAMPLE 17.6**

Determine the inverse of $A = \begin{pmatrix} 1 & -2 \\ 3 & 4 \end{pmatrix}$ and verify the result.

Now    $|A| = \begin{vmatrix} 1 & -2 \\ 3 & 4 \end{vmatrix} = (1 \times 4) - (3 \times -2) = 10$

Hence    $A^{-1} = \frac{1}{10} \begin{pmatrix} 4 & 2 \\ -3 & 1 \end{pmatrix} = \begin{pmatrix} 0.4 & 0.2 \\ -0.3 & 0.1 \end{pmatrix}$

To verify the result we have

$$AA^{-1} = \begin{pmatrix} 1 & -2 \\ 3 & 4 \end{pmatrix} \begin{pmatrix} 0.4 & 0.2 \\ -0.3 & 0.1 \end{pmatrix} = \begin{pmatrix} 1 & 0 \\ 0 & 1 \end{pmatrix} = I$$

## The property $AA^{-1} = A^{-1}A = I$

In Example 17.6 we have

$$A^{-1}A = \begin{pmatrix} 0.4 & 0.2 \\ -0.3 & 0.1 \end{pmatrix} \begin{pmatrix} 1 & -2 \\ 3 & 4 \end{pmatrix} = \begin{pmatrix} 1 & 0 \\ 0 & 1 \end{pmatrix} = I$$

and in general
$$\boxed{AA^{-1} = A^{-1}A = I}$$

# SINGULAR MATRIX

A matrix which does not have an inverse is called a *singular matrix*. This happens when $|A| = 0$

For example, since $\begin{vmatrix} 3 & 6 \\ 1 & 2 \end{vmatrix} = (3 \times 2) - (6 \times 1) = 0$

then $\begin{pmatrix} 3 & 6 \\ 1 & 2 \end{pmatrix}$ is a singular matrix.

## Exercise 17.4

Decide whether each of the matrices in Question 1-9 has an inverse. If the inverse exists, find it.

1) $\begin{pmatrix} 2 & 5 \\ 1 & 4 \end{pmatrix}$
2) $\begin{pmatrix} 2 & 5 \\ 1 & 3 \end{pmatrix}$
3) $\begin{pmatrix} 3 & 2 \\ 1 & 2 \end{pmatrix}$
4) $\begin{pmatrix} 4 & 10 \\ 2 & 5 \end{pmatrix}$

5) $\begin{pmatrix} 224 & 24 \\ 24 & 4 \end{pmatrix}$
6) $\begin{pmatrix} a & -b \\ -a & b \end{pmatrix}$
7) $\begin{pmatrix} 2 & 3 \\ -1 & 1 \end{pmatrix}$
8) $\begin{pmatrix} 2 & -3 \\ 1 & 5 \end{pmatrix}$

9) $\begin{pmatrix} 1 & 1 \\ 0 & 1 \end{pmatrix}$

10) Given that $A = \begin{pmatrix} 1 & 0 \\ 3 & 2 \end{pmatrix}$ and $B = \begin{pmatrix} 3 & 5 \\ 1 & 2 \end{pmatrix}$, calculate:

(a) $A^{-1}$         (b) $B^{-1}$         (c) $B^{-1}A^{-1}$         (d) $AB$

(e) $(AB)^{-1}$       (f) Compare the answers to (c) and (e)

# SYSTEMS OF LINEAR EQUATIONS

Given the system of equations $\left. \begin{matrix} 5x + y = 7 \\ 3x - 4y = 18 \end{matrix} \right\}$

we can rewrite it in the form $\begin{pmatrix} 5x + y \\ 3x - 4y \end{pmatrix} = \begin{pmatrix} 7 \\ 18 \end{pmatrix}$

or

$$\begin{pmatrix} 5 & 1 \\ 3 & -4 \end{pmatrix} \begin{pmatrix} x \\ y \end{pmatrix} = \begin{pmatrix} 7 \\ 18 \end{pmatrix}$$

That is

$$\begin{pmatrix} \text{Matrix of} \\ \text{coefficients} \end{pmatrix} \begin{pmatrix} \text{Matrix of} \\ \text{variables} \end{pmatrix} = \begin{pmatrix} \text{Matrix of} \\ \text{constants} \end{pmatrix}$$

Denote the matrix of coefficients by $C$ and its inverse by $C^{-1}$.

Then

$$|C| = \begin{vmatrix} 5 & 1 \\ 3 & -4 \end{vmatrix} = 5 \times (-4) - 1 \times 3 = -23$$

and

$$C^{-1} = \frac{1}{-23}\begin{pmatrix} -4 & -1 \\ -3 & 5 \end{pmatrix} = \frac{1}{23}\begin{pmatrix} 4 & 1 \\ 3 & -5 \end{pmatrix}$$

Now

$$C\begin{pmatrix} x \\ y \end{pmatrix} = \begin{pmatrix} 7 \\ 18 \end{pmatrix}$$

and multiplying both sides by $C^{-1}$ gives

$$C^{-1}C\begin{pmatrix} x \\ y \end{pmatrix} = C^{-1}\begin{pmatrix} 7 \\ 18 \end{pmatrix}$$

$$\therefore \quad I\begin{pmatrix} x \\ y \end{pmatrix} = C^{-1}\begin{pmatrix} 7 \\ 18 \end{pmatrix}$$

or

$$\begin{pmatrix} 1 & 0 \\ 0 & 1 \end{pmatrix} \times \begin{pmatrix} x \\ y \end{pmatrix} = \frac{1}{23}\begin{pmatrix} 4 & 1 \\ 3 & -5 \end{pmatrix} \times \begin{pmatrix} 7 \\ 18 \end{pmatrix}$$

$$\therefore \quad \begin{pmatrix} 1 \times x & 0 \times y \\ 0 \times x & 1 \times y \end{pmatrix} = \frac{1}{23}\begin{pmatrix} 4 \times 7 + & 1 \times 18 \\ 3 \times 7 + & (-5) \times 18 \end{pmatrix}$$

$$\therefore \quad \begin{pmatrix} x \\ y \end{pmatrix} = \frac{1}{23}\begin{pmatrix} 46 \\ -69 \end{pmatrix}$$

$$\therefore \quad \begin{pmatrix} x \\ y \end{pmatrix} = \begin{pmatrix} 2 \\ -3 \end{pmatrix}$$

Thus comparing the matrices shows that $x = 2$ and $y = -3$

We would not normally perform multiplication by the unit matrix.
We did so here to illustrate that when a matrix, here $\begin{pmatrix} x \\ y \end{pmatrix}$, is multiplied by the unit matrix then it is unaltered.

This confirms that the unit matrix performs as unity (the number one) in normal arithmetic.

# Exercise 17.5

Use matrix methods to solve each of the following systems of equations.

$$
\begin{aligned}
\text{1)} \quad & \left.\begin{aligned} x + y &= 1 \\ 3x + 2y &= 8 \end{aligned}\right\}
\end{aligned}
\qquad
\begin{aligned}
\text{2)} \quad & \left.\begin{aligned} x + y &= 6 \\ 3x - 2y &= -7 \end{aligned}\right\}
\end{aligned}
\qquad
\begin{aligned}
\text{3)} \quad & \left.\begin{aligned} 5x - 2y &= 17 \\ 2x + 3y &= 3 \end{aligned}\right\}
\end{aligned}
$$

$$
\begin{aligned}
\text{4)} \quad & \left.\begin{aligned} 3x - 2y &= 12 \\ 4x + y &= 5 \end{aligned}\right\}
\end{aligned}
\qquad
\begin{aligned}
\text{5)} \quad & \left.\begin{aligned} 3x + 2y &= 6 \\ 4x - y &= 5 \end{aligned}\right\}
\end{aligned}
\qquad
\begin{aligned}
\text{6)} \quad & \left.\begin{aligned} 3x - 4y &= 26 \\ 5x + 6y &= -20 \end{aligned}\right\}
\end{aligned}
$$

# 18. CALCULUS

After reaching the end of this chapter you should be able to:

1. Determine gradients of chord and tangent to a simple curve.
2. Deduce that the process of moving a point on a curve towards a fixed point on the curve causes the gradient of the chord joining the points to approach that of the tangent to the curve at the fixed point.
3. Identify incremental changes in x, y directions as $\delta x, \delta y$.
4. Determine the ratio $\dfrac{\delta y}{\delta x}$ in terms of x and $\delta x$ for the function in 1.
5. Derive the limit of $\dfrac{\delta y}{\delta x}$ as $\delta y$ tends to zero and defines it as $\dfrac{dy}{dx}$.
6. State that the rate of change at a maximum or minimum point of a curve is zero.
7. Derive $\dfrac{dy}{dx}$ for functions $y = ax^n$ for $n = 0$, 1, 2, and 3 from first principles.
8. Differentiate simple algebraic functions of the form $y = ax^n + bx^{n-1} + \ldots$
9. Demonstrate graphically results for the derivatives of $\sin \theta$ and $\cos \theta$
10. Differentiate functions of the form
$$y = a\cos \theta + b\sin \theta$$
11. Deduce the derivative of the exponential function, $e^x$.
12. Define indefinite integration as the reverse process of differentiation.
13. Determine indefinite integrals of simple functions.
14. Recognise the need to include an arbitrary constant of integration.
15. Define $\displaystyle\int_a^b y\, dx$ as the area under the curve between the ordinates $x = a$ and $x = b$.
16. Determine areas by applying the definite integral for simple functions.
17. Relate areas obtained by integration to results obtained by approximate methods.

## THE GRADIENT OF A CURVE

### Graphical Method

In mathematics and technology we often need to know the rate of change of one variable with respect to another. For instance, velocity is the rate of change of distance with respect to time, and acceleration is the rate of change of velocity with respect to time.

Consider the graph of $y = x^2$, part of which is shown in Fig. 18.1. As the values of $x$ increase so do the values of $y$, but they do not increase at the same rate. A glance at the portion of the curve shown shows that the values of $y$ increase faster when $x$ is large, because the gradient of the curve is increasing.

292

To find the rate of change of $y$ with respect to $x$ at a particular point we need to find the gradient of the curve at that point.

If we draw a tangent to the curve at the point, the gradient of the tangent will be the same as the gradient of the curve.

**EXAMPLE 18.1**

Find the gradient of the curve $y = x^2$ at the point where $x = 2$

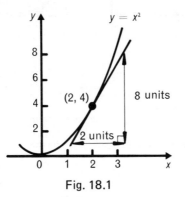

Fig. 18.1

The point where $x = 2$ is the point $(2, 4)$. We draw a tangent at this point, as shown in Fig. 18.1. Then by constructing a right-angled triangle the gradient is found to be $\dfrac{8}{2} = 4$. This gradient is positive, in accordance with our previous work, since the tangent slopes upwards from left to right.

**EXAMPLE 18.2**

Draw the graph of $y = x^2 - 3x + 7$ between $x = -4$ and $x = 3$ and hence find the gradient at: a) the point $x = -3$, b) the point $x = 2$.

a) At the point where $x = -3$,

$$y = (-3)^2 - 3(-3) + 7 = 25$$

At the point $(-3, 25)$ draw a tangent as shown in Fig. 18.2. The gradient is found by drawing a right-angled triangle (which should be as large as conveniently possible for accuracy) as shown, and measuring its height and base.

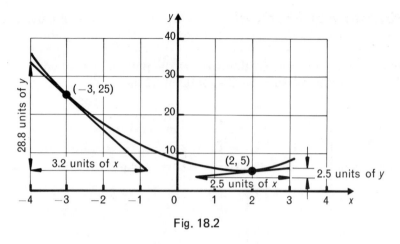

Fig. 18.2

Hence:    Gradient at point $(-3, 25)$ $= -\dfrac{28.8}{3.2} = -9$

the negative sign indicating a downward slope from left to right.

**b)** At the point where $x = 2$,
$$y = 2^2 - 3(2) + 7 = 5$$

Hence by drawing a tangent and a right-angled triangle at the point $(2, 5)$ in a similar manner to above,

$$\text{Gradient at point } (2, 5) = \frac{2.5}{2.5} = 1$$

being positive as the tangent slopes upwards from left to right.

## Exercise 18.1

1) Draw the graph of $2x^2 - 5$ for values of $x$ between $-2$ and $+3$. Draw, as accurately as possible, the tangents to the curve at the points where $x = -1$ and $x = +2$ and hence find the gradient of the curve at these points.

2) Draw the curve $y = x^2 - 3x + 2$ from $x = 2.5$ to $x = 3.5$ and find its gradient at the point where $x = 3$

3) Draw the curve $y = x - \dfrac{1}{x}$ from $x = 0.8$ to $1.2$. Find its gradient at $x = 1$

# Numerical Method

The gradient of a curve may always be found by graphical means but this method is often inconvenient. A numerical method will now be developed.

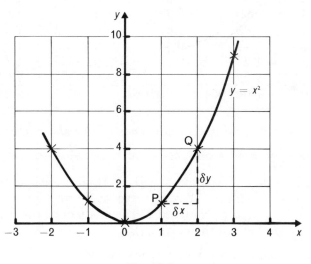

Fig. 18.3

Consider the curve $y = x^2$, part of this is shown in Fig. 18.3. Let P be the point on the curve at which $x = 1$ and $y = 1$. Q is a variable point on the curve, which will be considered to start at the point $(2, 4)$ and move down the curve towards P, rather like a bead slides down a wire.

The symbol $\delta x$ will be used to represent an increment of $x$, and $\delta y$ will be used to represent the corresponding increment of $y$. The gradient of the chord PQ is then $\dfrac{\delta y}{\delta x}$

When Q is at the point $(2, 4)$ then $\delta x = 1$, and $\delta y = 3$

$\therefore$ $$\frac{\delta y}{\delta x} = \frac{3}{1} = 3$$

The following table shows how $\dfrac{\delta y}{\delta x}$ alters as Q moves nearer and nearer to P.

| Co-ordinates of Q | | $\delta y$ | $\delta x$ | Gradient of $PQ = \dfrac{\delta y}{\delta x}$ |
|---|---|---|---|---|
| $x$ | $y$ | | | |
| 2 | 4 | 3 | 1 | 3 |
| 1.5 | 2.25 | 1.25 | 0.5 | 2.5 |
| 1.4 | 1.96 | 0.96 | 0.4 | 2.4 |
| 1.3 | 1.69 | 0.69 | 0.3 | 2.3 |
| 1.2 | 1.44 | 0.44 | 0.2 | 2.2 |
| 1.1 | 1.21 | 0.21 | 0.1 | 2.1 |
| 1.01 | 1.0201 | 0.0201 | 0.01 | 2.01 |
| 1.001 | 1.002001 | 0.002001 | 0.001 | 2.001 |

It will be seen that as Q approaches nearer and nearer to P, the value of $\dfrac{\delta y}{\delta x}$ approaches 2. It is reasonable to suppose that eventually when Q coincides with P (that is, when the chord PQ becomes a tangent to the curve at P) the gradient of the tangent will be exactly equal to 2. The gradient of the tangent will give us the gradient of the curve at P.

Now as Q approaches P, $\delta x$ tends to zero and the gradient of the chord, $\dfrac{\delta y}{\delta x}$, tends, in the limit (as we say), to the gradient of the tangent. We denote the gradient of the tangent as $\dfrac{dy}{dx}$. We can write all this as

$$\underset{\delta x \to 0}{\text{Limit}} \frac{\delta y}{\delta x} = \frac{dy}{dx}$$

Instead of selecting special values for $\delta y$ and $\delta x$ let us now consider the general case, so that P has the co-ordinates $(x, y)$ and Q has the co-ordinates $(x + \delta x, y + \delta y)$, (Fig. 18.4). Q is taken very close to P, so that $\delta x$ is a very small quantity.

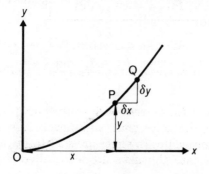

Fig. 18.4

Now $$y = x^2$$

and as $Q(x + \delta x, y + \delta y)$ lies on the curve, then

$$y + \delta y = (x + \delta x)^2$$

∴ $$y + \delta y = x^2 + 2x\delta x + (\delta x)^2$$

But $y = x^2$, so $$\delta y = 2x\delta x + (\delta x)^2$$

and, by dividing both sides by $\delta x$, the gradient of chord PQ is

$$\frac{\delta y}{\delta x} = 2x + \delta x$$

As Q approaches P, $\delta x$ tends to zero and $\dfrac{\delta y}{\delta x}$ tends, in the limit, to the gradient of the tangent of the curve at P.

Thus $$\underset{\delta x \to 0}{\text{Limit}} \frac{\delta y}{\delta x} = \frac{dy}{dx} = 2x$$

The process of finding $\dfrac{dy}{dx}$ is called *differentiation*.

The symbol $\dfrac{dy}{dx}$ means the differential coefficient of $y$ with respect to $x$.

We can now check our assumption regarding the gradient of the curve at P. Since at P the value of $x = 1$, then substituting in the expression $\dfrac{dy}{dx} = 2x$ we get $\dfrac{dy}{dx} = 2 \times 1 = 2$

and we see that our assumption was correct.

## TURNING POINTS

At the points P and Q (Fig. 18.5) the tangent to the curve is parallel to the $x$-axis. The points P and Q are called *turning points*. The turning point at P is called a *maximum* turning point and the turning point at Q is called a *minimum* turning point. It will be seen from Fig. 18.5 that the value of $y$ at P is not the greatest value of $y$ nor is the value of $y$ at Q the least. The terms maximum and minimum values apply only to the values of $y$ at the turning points and not to the values of $y$ in general.

In practical applications, however, we are usually concerned with a specific range of values of $x$ which are dictated by the problem. There is then no difficulty in identifying a particular maximum or minimum within this range of values of $x$.

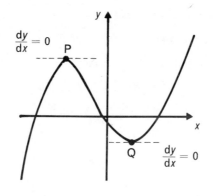

Fig. 18.5

Now the rate of change of $y$ with respect to $x$ at a particular point is given by the gradient of the tangent to the curve at that point. Thus at points P and Q the rate of change is zero or, in calculus notation, $\dfrac{dy}{dx} = 0$

# DIFFERENTIAL COEFFICIENT OF $x^n$

It can be shown, by a method similar to that used for finding the differential coefficient of $x^2$, that

| | |
|---|---|
| If | $y = x^n$ |
| then | $\dfrac{dy}{dx} = nx^{n-1}$ |

This is true for all values of $n$ including negative and fractional indices. When we use it as formula it enables us to avoid having to differentiate each time from first principles.

**EXAMPLE 18.3**

$$y = x^3$$
$$\therefore \quad \frac{dy}{dx} = 3x^2$$

$$y = \frac{1}{x} = x^{-1}$$
$$\therefore \quad \frac{dy}{dx} = -x^{-2} = -\frac{1}{x^2}$$

$$y = \sqrt{x} = x^{1/2}$$
$$\therefore \quad \frac{dy}{dx} = \frac{1}{2}x^{1/2} = \frac{1}{2}\frac{1}{x^{1/2}} = \frac{1}{2\sqrt{x}}$$

$$y = \sqrt[5]{x^2} = x^{2/5}$$
$$\therefore \quad \frac{dy}{dx} = \frac{2}{5}x^{2/5-1} = \frac{2}{5}x^{-3/5} = \frac{2}{5(\sqrt[5]{x^3})}$$

When a power of $x$ is multiplied by a constant, that constant remains unchanged by the process of differentiation.

| | |
|---|---|
| Hence if | $y = ax^n$ |
| then | $\dfrac{dy}{dx} = anx^{n-1}$ |

**EXAMPLE 18.4**

$y = 2x^{1.3}$

$\therefore \dfrac{dy}{dx} = 2(1.3)x^{0.3} = 2.6x^{0.3}$

$y = \tfrac{1}{5}x^7$

$\therefore \dfrac{dy}{dx} = \dfrac{1}{5} \times 7x^6 = \dfrac{7}{5}x^6$

---

$y = \tfrac{3}{4}\sqrt[3]{x} = \tfrac{3}{4}x^{1/3}$

$\therefore \dfrac{dy}{dx} = \dfrac{3}{4} \times \dfrac{1}{3}x^{-2/3} = \dfrac{1}{4}x^{-2/3}$

$y = \dfrac{4}{x^2} = 4x^{-2}$

$\therefore \dfrac{dy}{dx} = 4(-2)x^{-3} = -8x^{-3}$

**When a numerical constant is differentiated the result is zero.** This can be seen since $x^0 = 1$ and we can write, for example, constant 4 as $4x^0$, then differentiating with respect to $x$ we get

$$4(0)x^{-1} = 0$$

If, as an alternative method, we plot the graph of $y = 4$ we get a straight line parallel with the $x$-axis as shown in Fig. 18.6.

The gradient of the line is zero: that is, $\dfrac{dy}{dx} = 0$

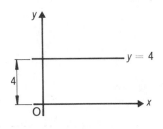

Fig. 18.6

To differentiate an expression containing the sum of several terms, differentiate each individual term separately.

**EXAMPLE 18.5**

a)    $y = 3x^2 + 2x + 3$

$\therefore$    $\dfrac{dy}{dx} = 3(2)x + 2(1)x^0 + 0 = 6x + 2$

b)    $y = ax^3 + bx^2 + cx + d$   where $a, b, c$ and $d$ are constants,

$\therefore$    $\dfrac{dy}{dx} = 3ax^2 + 2bx + c$

So far our differentiation has been in terms of $x$ and $y$ only. But they are only letters representing variables and we may choose other letters or symbols.

c)    $s = \sqrt{t} + \dfrac{1}{\sqrt{t}} = t^{1/2} + t^{-1/2}$

$\therefore$    $\dfrac{ds}{dt} = \dfrac{1}{2}t^{-1/2} + \left(-\dfrac{1}{2}\right)t^{-3/2} = \dfrac{1}{2\sqrt{t}} - \dfrac{1}{2\sqrt{t^3}}$

d)    $v = 3.1u^{1.4} - \dfrac{3}{u} + 5 = 3.1u^{1.4} - 3u^{-1} + 5$

$\therefore$    $\dfrac{dv}{du} = (3.1)(1.4)u^{0.4} - 3(-1)u^{-2} = 4.34u^{0.4} + \dfrac{3}{u^2}$

# Finding the Gradient of a Curve by Differentiation

**EXAMPLE 18.6**

Find the gradient of the graph $y = 3x^2 - 3x + 4$:

a) when $x = 3$, and,      b) when $x = -2$

The gradient at a point is expressed by $\dfrac{dy}{dx}$

If           $y = 3x^2 - 3x + 4$

then        $\dfrac{dy}{dx} = 6x - 3$

a) When   $x = 3$                    b) When   $x = -2$

$\dfrac{dy}{dx} = 6(3) - 3 = 15$            $\dfrac{dy}{dx} = 6(-2) - 3 = -15$

# Exercise 18.2

Differentiate the following:

1) $y = x^2$                    2) $y = x^7$                    3) $y = 4x^3$

4) $y = 6x^5$                  5) $s = 0.5t^3$                6) $A = \pi R^2$

7) $y = x^{1/2}$                8) $y = 4x^{3/2}$              9) $y = 2\sqrt{x}$

10) $y = 3 \times \sqrt[3]{x^2}$        11) $y = \dfrac{1}{x^2}$            12) $y = \dfrac{1}{x}$

13) $y = \dfrac{3}{5x}$              14) $y = \dfrac{2}{x^3}$              15) $y = \dfrac{1}{\sqrt{x}}$

16) $y = \dfrac{2}{3\sqrt{x}}$            17) $y = \dfrac{5}{x\sqrt{x}}$            18) $s = \dfrac{3\sqrt{t}}{5}$

19) $K = \dfrac{0.01}{h}$                        20) $y = \dfrac{5}{x^7}$

21) $y = 4x^2 - 3x + 2$                    22) $s = 3t^3 - 2t^2 + 5t - 3$

23) $q = 2u^2 - u + 7$                      24) $y = 5x^4 - 7x^3 + 3x^2 - 2x + 5$

25) $s = 7t^5 - 3t^2 + 7$                    26) $y = \dfrac{x + x^3}{\sqrt{x}}$

27) $y = \dfrac{3 + x^2}{x}$          28) $y = \sqrt{x} + \dfrac{1}{\sqrt{x}}$        29) $y = x^3 + \dfrac{3}{\sqrt{x}}$

30) $s = t^{1.3} - \dfrac{1}{4t^{2.3}}$                  31) $y = \dfrac{3x^3}{5} - \dfrac{2x^2}{7} + \sqrt{x}$

32) $y = 0.08 + \dfrac{0.01}{x}$                  33) $y = 3.1x^{1.5} - 2.4x^{0.6}$

34) $y = \dfrac{x^3}{2} - \dfrac{5}{x} + 3$                  35) $s = 10 - 6t + 7t^2 - 2t^3$

36) Find the gradient of the curve $y = 3x^2 + 7x + 3$ at the points where $x = -2$ and $x = 2$

37) Find the gradient of the curve $y = 2x^3 - 7x^2 + 5x - 3$ at the points where $x = -1.5$, $x = 0$ and $x = 3$

38) Find the values of $x$ for which the gradient of the curve $y = 3 + 4x - x^2$ is equal to:
(a) $-1$      (b) $0$      (c) $2$?

# TO FIND $\dfrac{d}{d\theta}$ (sin $\theta$)

# OR THE RATE OF CHANGE OF sin$\theta$

The rate of change of a curve at any point is the gradient of the tangent at that point. We shall, therefore, find the gradient at various points on the graph of sin $\theta$ and then plot the values of these gradients to obtain a new graph.

It is suggested that the reader follows the method given, plotting his own curves on graph paper.

First, we plot the graph of $y = \sin\theta$ from $\theta = 0°$ to $\theta = 90°$ using values of sin $\theta$ obtained from tables which are:

| $\theta$ | 0° | 15° | 30° | 45° | 60° | 75° | 90° |
|---|---|---|---|---|---|---|---|
| $y = \sin\theta$ | 0 | 0.259 | 0.500 | 0.707 | 0.866 | 0.966 | 1.000 |

These values are shown plotted in Fig. 18.7.

Consider point P on the curve, where $\theta = 45°$, and draw the tangent APM.

We can find the gradient of the tangent by constructing a suitable right-angled triangle AMN (which should be as large as conveniently possible for accuracy) and finding the value of $\dfrac{MN}{AN}$.

Using the scale on the $y$-axis gives, by measurement, MN = 1.29.
Using the scale on the $\theta$-axis gives, by measurement, AN = 104°.

In calculations of this type it is necessary to obtain AN in radians.

Since $\qquad\qquad 360° = 2\pi$ radians

then $\qquad\qquad\qquad 1° = \dfrac{2\pi}{360}$ radians

and $\qquad\qquad 104° = \dfrac{2\pi}{360} \times 104 = 1.81$ radians

Hence $\qquad$ Gradient at P $= \dfrac{MN}{AN} = \dfrac{1.29}{1.81} = 0.71$

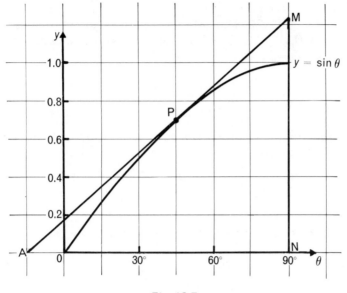

Fig. 18.7

The value 0.71 is used as the y-value at $\theta = 45°$ to plot a point on a new graph using the same scales as before. This new graph could be plotted on the same axes as $y = \sin\theta$ but for clarity it has been shown on new axes in Fig. 18.8.

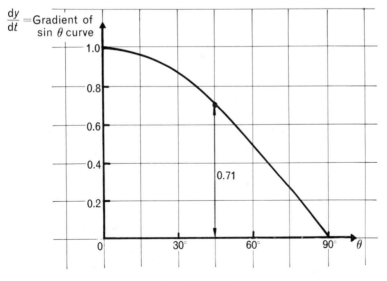

Fig. 18.8

This procedure is repeated for points on the sin $\theta$ curve at $\theta$ values $0°$, $15°$, $30°$, $60°$, $75°$ and $90°$ and the new curve obtained will be as shown in Fig. 18.8. This is the graph of the gradients of the sine curve at various points.

If we now plot a graph of $\cos \theta$, taking values from tables, on the axes in Fig. 18.8 we shall find that the two curves coincide—any difference will be due to errors from drawing the tangents.

Hence the gradient of the sin $\theta$ curve at any value of $\theta$ is the same as the value of $\cos \theta$.

In other words the rate of change of sin $\theta$ is $\cos \theta$, provided that the angle $\theta$ is in radians.

In the above work we have only considered the graphs between $0°$ and $90°$ but the results are true for all values of the angle.

Hence if $\qquad y = \sin \theta \quad$ then $\quad \dfrac{dy}{dx} = \cos \theta$

or $\qquad \boxed{\dfrac{d}{d\theta}\left(\sin \theta\right) = \cos \theta} \qquad$ provided that $\theta$ is in radians

The same procedure may be used to show that

$$\boxed{\dfrac{d}{d\theta}\left(\cos \theta\right) = -\sin \theta} \qquad \text{provided that } \theta \text{ is in radians.}$$

**EXAMPLE 18.7**

Find $\dfrac{dy}{dx}$ if $y = 3 \sin x + 2 \cos x$

We have $\qquad y = \sin x + \sin x + \sin x + \cos x + \cos x$

Since this is a sum of terms we may differentiate each in turn.

Thus $\qquad \dfrac{dy}{dx} = \cos x + \cos x + \cos x + (-\sin x) + (-\sin x)$

$\therefore \qquad \dfrac{dy}{dx} = 3 \cos x - 2 \sin x$

We can see from the above result that when either $\sin x$ or $\cos x$ are preceded by a constant multiplier it does not affect the differentiation but merely remains there.

**EXAMPLE 18.8**

Find the value of $\dfrac{d}{d\theta}(5 \sin \theta + 3 \cos \theta)$ if $\theta$ has a value equivalent to $25°$.

Let $\qquad y = 5 \sin \theta + 3 \cos \theta$

$\therefore \qquad \dfrac{dy}{d\theta} = 5 \cos \theta + 3(-\sin \theta)$

$\qquad\qquad = 5 \cos \theta - 3 \sin \theta$

Thus when $\theta = 25°$

$\qquad \dfrac{dy}{d\theta} = 5 \cos 25° - 3 \sin 25°$

A suitable sequence of operation on the calculating machine is:

Set the machine for calculations in degrees:

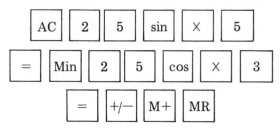

giving an answer 3.264 correct to four significant figures.

# TO FIND $\dfrac{d}{dx}(e^x)$ OR THE RATE OF CHANGE OF $e^x$

You may remember from p. 254 the important property of the exponential function, $e^x$ that:

At any point of the curve of $e^x$ the gradient of the curve is equal to the value of $e^x$.

In the calculus notation this may be stated as

$$\frac{d}{dx}(e^x) = e^x$$

EXAMPLE 18.9

Find the value of $\dfrac{d}{dx}(4e^x)$ when $x$ has a value of 2.32

Let                                      $y = 4e^x$

then                                     $\dfrac{dy}{dx} = 4e^x$

Thus when $x = 2.32$:          $\dfrac{dy}{dx} = 4e^{2.32}$

A suitable sequence of operations on the calculating machine is:

| AC | 2 | . | 3 | 2 | $e^x$ | × | 4 | = |

giving an answer 40.70 correct to four significant figures.

## Exercise 18.3

1) Draw the curve of $\cos\theta$ between $\theta = 0°$ and $90°$ and show by graphical differentiation that the differential coefficient of $\cos\theta$ is $-\sin\theta$.

2) If $y = 3\cos\theta - 2\sin\theta$, find the value of $\dfrac{dy}{d\theta}$ if $\theta$ has a value equivalent to $20°$.

3) If $x = 0.22$, find the value of $\dfrac{d}{dx}(6e^x)$

4) Find $\dfrac{dy}{dx}$ if $y = 5\sin x + 3\cos x$

5) Find $\dfrac{d}{d\theta}(4\sin\theta - 2\cos\theta)$

## INTEGRATION

## Integration as the Inverse of Differentiation

We have previously discovered how to obtain the differential coefficients of various functions. Our objective in this section is to find out how to reverse the process. That is, being given the differential coefficient of a function we try to discover the original function.

If $$y = \frac{x^4}{4}$$

then $$\frac{dy}{dx} = x^3$$

or we may write $$dy = x^3\,dx$$

The expression $x^3\,dx$ is called the differential of $\dfrac{x^4}{4}$

Reversing the process of differentiation is called *integration*.

It is indicated by using the integration sign $\int$ in front of the differential.

Thus, if: $$dy = x^3\,dx$$

then reversing the process $$y = \int x^3\,dx = \frac{x^4}{4}$$

Similarly if $$y = \frac{x^5}{5}$$

then $$\frac{dy}{dx} = x^4$$

or $$dy = x^4\,dx$$

and reversing the process $$y = \int x^4\,dx = \frac{x^5}{5}$$

Also, if $$y = \frac{x^{n+1}}{n+1}$$

then $$\frac{dy}{dx} = x^n$$

or $$dy = x^n\,dx$$

from which $$y = \int x^n\,dx = \frac{x^{n+1}}{n+1}$$

Now $\dfrac{x^{n+1}}{n+1}$ is called the integral of $x^n\,dx$

This rule applies to all indices, positive, negative and fractional except for $\int x^{-1}\,dx$, which is a special case beyond the scope of this book.

Since the differential coefficient of $\sin x$ is $\cos x$ it follows that the integral of $\cos x$, with respect to $x$, is $\sin x$

Similarly the integral of $\sin x$, with respect to $x$, is $-\cos x$

Also the integral, of $e^x$ with respect to $x$ is $e^x$

Summarising all these results we have:

| $\int x^n\,dx \;=\; \dfrac{x^{n+1}}{n+1}$ | $\int \sin x\,dx \;=\; -\cos x$ |
|---|---|
| $\int e^x\,dx \;=\; e^x$ | $\int \cos x\,dx \;=\; \sin x$ |

## THE CONSTANT OF INTEGRATION

We know that the differential of $\dfrac{x^2}{2}$ is $x\,dx$. Therefore if we are asked to integrate $x\,dx$, $\dfrac{x^2}{2}$ is one answer; but it is not the only possible answer because $\dfrac{x^2}{2}+2$, $\dfrac{x^2}{2}+5$, $\dfrac{x^2}{2}+19$, etc. are all expressions whose differential is $x\,dx$. The general expression for $\int x\,dx$ is therefore $\dfrac{x^2}{2}+c$, where $c$ is a constant known as the constant of integration. Each time we integrate the constant of integration must be added.

**EXAMPLE 18.10**

a) $\int x^5\,dx = \dfrac{x^{5+1}}{5+1}+c = \dfrac{x^6}{6}+c$ 

b) $\int x\,dx = \dfrac{x^{1+1}}{1+1}+c = \dfrac{x^2}{2}+c$

c) $\int \sqrt{x}\,dx = \int x^{1/2}\,dx = \dfrac{x^{3/2}}{3/2}+c = \dfrac{2x^{3/2}}{3}+c$

d) $\int \dfrac{dx}{x^3} = \int x^{-3}\,dx = \dfrac{x^{-2}}{-2}+c = -\dfrac{1}{2x^2}+c$

e) $\int \cos x\,dx = \sin x + c$

# A Constant Coefficient May be Taken Outside the Integral Sign

**EXAMPLE 18.11**

a) $\int 3x^2 \, dx = 3 \int x^2 \, dx = 3\dfrac{x^3}{3} + c = x^3 + c$

b) $\int 4 \sin \theta \, d\theta = 4 \int \sin \theta \, d\theta = 4(-\cos \theta) + c = -4 \cos \theta + c$

# The Integral of a Sum is the Sum of Their Separate Integrals

**EXAMPLE 18.12**

a) $\int (x^2 + x) \, dx$

Integrate each term separately:

$$\int x^2 \, dx = \frac{x^3}{3} \quad \text{and} \quad \int x \, dx = \frac{x^2}{2}$$

Thus $\int (x^2 + x) \, dx = \dfrac{x^3}{3} + \dfrac{x^2}{2} + c$

b) $\int (3x^4 + e^x - 6) \, dx = \frac{3}{5}x^5 + e^x - 6x + c$

c) $\int (3 \sin t - 5 \cos t) \, dt = -3 \cos t - 5 \sin t + c$

d) $\int (2x + 5)^2 \, dx = \int (4x^2 + 20x + 25) \, dx = \dfrac{4x^3}{3} + 10x^2 + 25x + c$

# Exercise 18.4

Integrate with respect to $x$:

1) $x^2$

2) $x^8$

3) $\sqrt{x}$

4) $\dfrac{1}{x^2}$

5) $\dfrac{1}{x^4}$

6) $\dfrac{1}{\sqrt{x}}$

7) $3x^4$

8) $5x^8 + e^x$

9) $x^2 + x + 3$

10) $2x^3 - 7x - 4$

11) $x^2 - 5x + \dfrac{1}{\sqrt{x}} + \dfrac{2}{x^2}$

12) $\dfrac{8}{x^3} - \dfrac{2}{x^2} + \sqrt{x}$     13) $(x-2)(x-1)$     14) $(x+3)^2$

15) $(2x-7)^2$     16) $2\cos x + 3\sin x$

## Evaluating the Constant of Integration

The value of the constant of integration may be found provided a corresponding pair of values of $x$ and $y$ are known.

### EXAMPLE 18.13

The gradient of the curve which passes through the point $(2, 3)$ is given by $x^2$. Find the equation of the curve.

We are given $\qquad\qquad\qquad \dfrac{dy}{dx} = x^2$

$\therefore \qquad\qquad\qquad\qquad y = \int x^2 \, dx = \dfrac{x^3}{3} + c$

We are also given that when $\qquad x = 2, \quad y = 3$

Substituting these values in $\qquad y = \dfrac{x^3}{3} + c$

we have $\qquad\qquad\qquad\qquad 3 = \dfrac{2^3}{3} + c$

$\therefore \qquad\qquad\qquad\qquad c = \dfrac{1}{3}$

Hence the equation of the curve is

$$y = \dfrac{x^3}{3} + \dfrac{1}{3}$$

or $\qquad\qquad\qquad\qquad y = \dfrac{1}{3}(x^3 + 1)$

## Exercise 18.5

1) The gradient of the curve which passes through the point $(2, 3)$ is given by $x$. Find the equation of the curve.

2) The gradient of the curve which passes through the point $(3, 8)$ is given by $(x^2 + 3)$. Find the value of $y$ when $x = 5$

**3)** It is known that for a certain curve $\dfrac{dy}{dx} = 3 - 2x$ and the curve cuts the $x$-axis where $x = 5$. Express $y$ in terms of $x$. State the length of the intercept on the $y$-axis.

**4)** Find the equation of the curve which passes through the point $(1, 4)$ and is such that $\dfrac{dy}{dx} = 2x^2 + 3x + 2$

**5)** If $\dfrac{dp}{dt} = (3 - t)^2$ find $p$ in terms of $t$ given that $p = 3$ when $t = 2$

**6)** The gradient of a curve is $ax + b$ at all points, where $a$ and $b$ are constants. Find the equation of the curve given that it passes through the points $(0, 4)$ and $(1, 3)$ and that the tangent at $(1, 3)$ is parallel to the $x$-axis.

**7)** Find the equation of the curve which passes through the point $(1, 2)$ and has the property of $\dfrac{dy}{dx} = e^x$

**8)** A curve is such that $\dfrac{dy}{d\theta} = \cos\theta$, and also $y = 1$ when $\theta = \dfrac{\pi}{2}$ radians. Find the equation of the curve.

**9)** At any point on a curve $\dfrac{dy}{dt} = 3\sin t$. Find the equation of the curve given that $y = 2$ when $t$ has a value equivalent to $25$ degrees.

## THE DEFINITE INTEGRAL

It has been shown that $\displaystyle\int x^n \, dx = \dfrac{x^{n+1}}{n+1} + c$

Since the expression contains an arbitrary constant $c$, the value of which is not known, it is called an indefinite integral.

A definite integral has a specific numerical answer without an unknown constant. An example of a definite integral is $\displaystyle\int_a^b x^n \, dx$.

$a$ and $b$ are called limits, $a$ being the lower limit and $b$ the upper limit. The method of evaluating a definite integral is shown in the following examples.

**EXAMPLE 18.14**

Find the value of $\displaystyle\int_2^3 x^2\,dx$

$$\int_2^3 x^2\,dx \;=\; \left[\frac{x^3}{3}+c\right]_2^3$$

$$=\; \left(\text{Value of } \frac{x^3}{3}+c \text{ when } x \text{ is put equal to } 3\right)$$

$$-\left(\text{Value of } \frac{x^3}{3}+c \text{ when } x \text{ is put equal to } 2\right)$$

$$=\; \left(\frac{3^3}{3}+c\right)-\left(\frac{2^3}{3}+c\right)$$

$$=\; \frac{27}{3}+c-\frac{8}{3}-c$$

$$=\; \frac{19}{3}\;=\;6.33$$

In integration the use of the square brackets, as in the above solution, has a specific meaning, that is *the integration of each term has been completed and the next step is to substitute the values of the limits for x.*

We should also note that the constant $c$ cancelled out. This will always happen and in solving definite integrals it is usual to omit $c$ as shown in the next example.

**EXAMPLE 18.15**

Find the value of $\displaystyle\int_1^2 (3x^2-2x+5)\,dx$

$$\int_1^2 (3x^2-2x+5) \;=\; \left[x^3-x^2+5x\right]_1^2$$

$$=\; (2^3-2^2+5\times 2)-(1^3-1^2+5\times 1)$$

$$=\; 14-5\;=\;9$$

**EXAMPLE 18.16**

Find the value of $\displaystyle\int_0^{\pi/2} \sin\theta\,d\theta$

$$\int_0^{\pi/2} \sin\theta\,d\theta \;=\; \left[-\cos\theta\right]_0^{\pi/2} \;=\; \left(-\cos\frac{\pi}{2}\right)-\left(-\cos 0\right)$$

$$=\; 0-(-1)\;=\;1$$

# Exercise 18.6

Evaluate the following definite integrals:

1) $\displaystyle\int_{1}^{2} x^2 \, dx$

2) $\displaystyle\int_{2}^{3} (2x+3) \, dx$

3) $\displaystyle\int_{0}^{2} (x^2+3) \, dx$

4) $\displaystyle\int_{1}^{2} (3x^2-4x+3) \, dx$

5) $\displaystyle\int_{1}^{2} x(2x-1) \, dx$

6) $\displaystyle\int_{0}^{2} \sqrt{x} \, dx$

7) $\displaystyle\int_{1}^{3} \frac{1}{x^2} \, dx$

8) $\displaystyle\int_{2}^{4} (x-1)(x-3) \, dx$

9) $\displaystyle\int_{0}^{\pi/2} \cos\phi \, d\phi$

10) $\displaystyle\int_{1}^{3} e^t \, dt$

# AREA UNDER A CURVE

Suppose that we wish to find the shaded area shown in Fig. 18.9. P, whose co-ordinates are $(x, y)$ is a point on the curve.

Let us now draw, below P, a vertical strip whose width $\delta x$ is very small. Since the width of the strip is very small we may consider

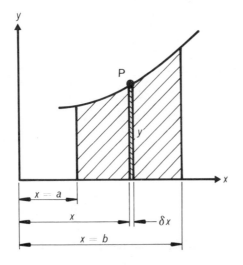

Fig. 18.9

the strip to be a rectangle with height $y$. Hence the area of the strip is approximately $y \times \delta x$. Such a strip is called an elementary strip and we will consider that the shaded area is made up from many elementary strips. Hence the required area is the sum of all the elementary strip areas between the values $x = a$ and $x = b$. In mathematical notation this may be stated as:

$$\text{Area} = \sum_{x=a}^{x=b} y \times \delta x \quad \text{approximately}$$

The process of integration may be considered to sum up an infinite number of elementary strips and hence gives an exact result.

$$\therefore \qquad \text{Area} = \int_{a}^{b} y \, dx \quad \text{exactly}$$

### EXAMPLE 18.17

Find the area bounded by the curve $y = x^3 + 3$, the $x$-axis and the lines $x = 1$ and $x = 3$

It is always wise to sketch the graph of given curve and show the area required together with an elementary strip, as shown in Fig. 18.10.

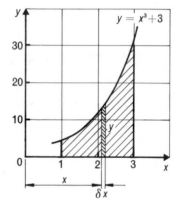

Fig. 18.10

The required area $= \sum_{x=1}^{x=3} y \times \delta x \quad$ approximately

$$= \int_{1}^{3} y \, dx \quad \text{exactly}$$

$$= \int_{1}^{3} (x^3 + 3) \, dx$$

$$= \left[ \frac{x^4}{4} + 3x \right]_{1}^{3} = \left( \frac{3^4}{4} + 3 \times 3 \right) - \left( \frac{1^4}{4} + 3 \times 1 \right)$$

$$= 26 \text{ square units}$$

## EXAMPLE 18.18

Find the area under the curve of $2\cos\theta$ between $\theta = 20°$ and $\theta = 60°$.

The curve of $2\cos\theta$ is shown in Fig. 18.11 from $0°$ to $90°$ and the required area together with an elementary strip.

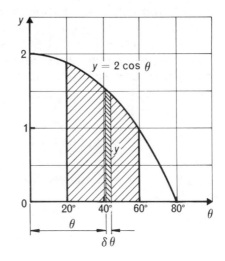

Fig. 18.11

The required area $= \displaystyle\sum_{\theta=20°}^{\theta=60°} y \times \delta\theta$ approximately

$$= 2\int_{20°}^{60°} \cos\theta \, d\theta \quad \text{exactly}$$

$$= 2\Big[\sin\theta\Big]_{20°}^{60°} = 2\Big(\sin 60° - \sin 20°\Big)$$

$$= 1.05 \text{ square units}$$

# Exercise 18.7

1) Find the area between the curve $y = x^3$, the $x$-axis and the lines $x = 5$ and $x = 3$

2) Find the area between the curve $y = 3 + 2x + 3x^2$, the $x$-axis and the lines $x = 1$ and $x = 4$

3) Find the area between the curve $y = x^2(2x - 1)$, the $x$-axis and the lines $x = 1$ and $x = 2$

4) Find the area between the curve $y = \dfrac{1}{x^2}$, the $x$-axis and the lines $x = 1$ and $x = 3$

5) Find the area between the curve $y = 5x - x^3$, the $x$-axis and the lines $x = 1$ and $x = 2$

6) Evaluate the integral $\int_0^{2\pi} \sin\theta \, d\theta$ and explain the result with reference to a sketched graph.

7) Find the area under the curve $2\sin\theta + 3\cos\theta$ between $\theta = 0$ and $\theta = \pi$ radians.

8) Find the area under the curve of $y = \sin\phi$ between $\phi = 0$ and $\phi = \pi$ radians.

9) Find the area under the curve $y = e^x$ between the ordinates $x = 0$ and $x = 2$

10) Find the area under the curve $y = 5e^x$ from $x = -0.5$ to $x = +0.5$

# ANSWERS

## ANSWERS TO CHAPTER 1

### Exercise 1.1

| | |
|---|---|
| 1) 13.143 | 2) $-11.35$ |
| 3) 27.4 | 4) 0.001 49 |
| 5) 1.94 | 6) $-4.26$ |
| 7) 1.28 | 8) 18.8 |
| 9) $-2.52$ | 10) 527 |
| 11) $-22.8$ | 12) $-22.8$ |
| 13) 0.007 58 | 14) $-0.348$ |
| 15) 0.657 | 16) 0.549 |
| 17) $-4.07$ | 18) 3.96 |
| 19) 6.61 | 20) 0.001 29 |

### Exercise 1.2

| | |
|---|---|
| 1) 29.0 | 2) 12.3 |
| 3) 0.0391 | 4) 0.0160 |
| 5) 0.0103 | 6) 94.2 |
| 7) 42.1 | 8) 86.3 |
| 9) 0.670 | 10) 2.90 |
| 11) 2.54 | 12) 0.506 |
| 13) 17.1 | 14) 2.84 |
| 15) 1.62 | 16) 0.518 |
| 17) 468 | 18) 3.22 |
| 19) 62.6 | 20) $1.92 \times 10^8$ |
| 21) 1.18 | 22) 9.19 |
| 23) 0.0171 | 24) $6.56 \times 10^6$ |
| 25) 5500 | 26) 614 |
| 27) 70.3 | 28) 55.8° |
| 29) 45.2° | 30) 18.6 |
| 31) (a) 72.4 | (b) 3.22   (c) 244 |
| (d) 10.7 | |

## ANSWERS TO CHAPTER 2

### Exercise 2.1

1) 21, 9, 0, $-6$, $-9$, $-6$, 0, 9
2) 4.20, 7.15, 12.87, 34.03,
   $-68.58$, $-16.48$, $-8.57$

3) 5.00, 3.69, 2.15, 0.29, $-2.04$,
   $-5.13$, $-9.62$
4) 123, 112, 79.8, 37.0, $-6.04$,
   $-40.3$, $-61.6$, $-71.3$
5) £22.63

## ANSWERS TO CHAPTER 3

### EXERCISE 3.1

1) $T = \dfrac{PV}{R}$     2) $T = \dfrac{ts}{S}$

3) $u = v - at$     4) $r = \dfrac{n-p}{c}$

5) $C = \frac{5}{9}(F - 32)$   6) $l = \dfrac{H-s}{q}$

7) $a = \dfrac{115 - S}{13}$   8) $r = \dfrac{R(V-2)}{V}$

9) $R = \dfrac{E - Cr}{C}$   10) $h = \dfrac{s}{\pi r} - r$

11) $T = \dfrac{H}{ws} + t$   12) $N = 2Cp + n$

13) $a = \dfrac{S}{n} - \dfrac{d(n-1)}{2}$

14) $R_1 = \dfrac{R_2 R}{R_2 - R}$

15) $r = \dfrac{c}{\pi h^2} + \dfrac{h}{3}$

16) $R = \dfrac{Vr}{V-2}$   17) $C = \dfrac{SF}{S-P}$

18) $x = \dfrac{2-y}{1+y}$   19) $x = \dfrac{p}{2-q}$

20) $u = \dfrac{T + 5T^2}{2 - 5T}$

### Exercise 3.2

1) $h = \dfrac{v^2}{2g}$     2) $r = \sqrt{\dfrac{A}{\pi}}$

$$3)\ v = \sqrt{\frac{2E}{m}} \qquad 4)\ d = \frac{w^2}{100}$$

$$5)\ l = g/(2\pi f)^2 \qquad 6)\ c = \sqrt{a^2 - b^2}$$

$$7)\ L = \frac{D^2}{1.44d} \qquad 8)\ y = \sqrt{z^2 - x^2}$$

$$9)\ f = \sqrt{\frac{2EP}{V}} \qquad 10)\ r = \sqrt{\frac{8vln}{\pi p}}$$

$$11)\ r = \frac{c^2 + 4h^2}{8h}$$

$$12)\ R = \sqrt{\frac{1}{8}\left(\frac{S}{\pi r}\right)^2 - r^2}$$

$$13)\ C = \frac{1}{L(2\pi f)^2}$$

$$14)\ v = \sqrt{2(E - mgh)/m}$$

$$15)\ l = (V - \pi r^2)/\pi r$$

$$16)\ P = \frac{f(D^2 - d^2)}{D^2 + d^2}$$

## ANSWERS TO CHAPTER 4

### Exercise 4.1

1) $m = 1, c = 3$
2) (a) $m = 1, c = 3$
   (b) $m = -3, c = 4$
   (c) $m = -3.1, c = -1.7$
   (d) $m = 4.3, c = -2.5$
3) $m = 2, c = 1$
4) $a = 0.25, b = 1.25$
5) $a = 0.29, b = -1.0$
6) 529 N
7) $E = 0.0984W + 0.72$
8) $a = 0.03, b = 0$
9) $a = 100, b = 0.43$
10) 524 N/m      11) 51 ohms

## ANSWERS TO CHAPTER 5

### Exercise 5.1

1) 1, 2      2) 4, 5      3) 4, 1
4) 40, 75    5) 8.9, 7.3
6) 25 ohms, 0.005, 31.25 ohms
7) £0.10, £0.30    8) £48, £150
9) £0.90, £1.25    10) 0.4, 50, 170 N
11) £250, £6500

12) £14 000, £16 000
13) 2, 8            14) 18, 28
15) $x = 5400$,
    $y = 1700$; £5400, £3400
16) 3, 8

### Exercise 5.2

1) 3, 4      2) 4 repeated
3) $+3, -3$  4) 3, $-5$    5) 0.667, 7
6) $-5, -1.5$    7) 2.414, $-0.414$
8) 2.181, 0.153    9) $+0.745, -0.745$

### Exercise 5.3

1) $x = -1, y = 1; x = 4, y = 6$
2) $x = 0, y = 5; x = 3, y = 11$
3) $x = 1, y = 3; x = -0.2, y = -3$
4) $x = 2.39, y = 6.91$;
   $x = 0.26, y = 0.54$
5) 60 m × 80 m

### Exercise 5.4

1) 1, $-1$, 4      2) 2, 0.33, $-1.20$
3) 1                4) $-1$, 1(repeated)
5) 3, 1.18, $-0.43$
6) 2                7) 1, 0.57, $-2.91$
8) $-3$, 2(repeated)
9) 2, 0.21, $-1.35$
10) $-1$            11) 7.9 m
12) 3.32 m          13) 5 m
14) 3

## ANSWERS TO CHAPTER 6

### Exercise 6.1

1) (a) 0.6109      (b) 1.457
   (c) 0.3367      (d) 0.7621
2) (a) 9°55'25"    (b) 89°33'53"
   (c) 4°29'11"
3) (a) 1.05 m      (b) 22.9 mm
4) (a) 120°        (b) 10.2°
5) 89.2 mm
6) (a) 4.71 m²     (b) 508 mm²
   (c) 7620 mm²
7) 866 mm²
8) 29.3 and 80.7 mm
9) 1240 mm²    10) 185 mm²
11) 163 mm²
12) 369 mm, 20 600 mm²
13) 7.96 m, 31.7 m²
14) 11.2 m²        15) 17.9 m

## ANSWERS TO CHAPTER 7

### Exercise 7.1

| | |
|---|---|
| 1) 67.9 mm | 2) 92.2 mm |
| 3) 115 mm | 4) 167 mm |
| 5) 39.7 mm | 6) 255 mm |
| 7) $30°33'$ | 8) $21°33'$ |
| 9) $64°45'$ | 10) 177 mm |

### Exercise 7.2

1) (a) $0.8572, -0.5150, -1.6643$
 (b) $0.0282, -0.9996, -0.0282$
 (c) $0.9764, -0.2162, -4.5169$
 (d) $-0.5150, -0.8572, 0.6009$
 (e) $-0.8597, -0.5108, 1.6831$
 (f) $-0.9798, -0.2000, 4.9006$
 (g) $-0.6633, 0.7484, -0.8863$
 (h) $-0.8887, 0.4584, -1.9389$
2) 1.897     3) 3.0248
4) $-0.28$     5) $-\frac{4}{5}, -\frac{3}{4}$
6) $8°14', 171°46'$
7) $153°13', 206°47'$
8) (a) $45°32', 134°28'$
 (b) $118°46'$   (c) $43°28'$
 (d) $119°33'$
9) (a) $232°$ and $308°$
 (b) $304°$     (c) $289°$
10) $14°34', 165°26'$
11) $105°42', 254°18'$

### Exercise 7.3

1) (a) $0.6157, 0.7880$
 (b) $0.9551, 0.3090$
 (c) $0.6157, -0.7880$
 (d) $0.9551, -0.3090$
 (e) $-0.3420, -0.9397$
 (f) $-0.9397, -0.3420$
 (g) $-0.8192, 0.5736$
 (h) $-0.5299, 0.8480$
2) $30°, 150°, -2.82$
3) $78°, 282°$     4) $23°, 157°$

### Exercise 7.4

1) (a) $C = 71°$, $b = 59.1$ mm,
 $c = 99.9$ mm
 (b) $A = 48°$, $a = 71.5$ mm,
 $c = 84.2$ mm
 (c) $B = 56°$, $a = 3.74$ m,
 $b = 9.53$ m
 (d) $B = 46°$, $b = 136$ mm,
 $c = 58.4$ mm

(e) $C = 67°$, $a = 1.51$ m,
 $c = 2.36$ m
(f) $C = 60°32'$,
 $a = 9.486$ mm
 $b = 11.56$ mm
(g) $B = 135°38'$,
 $a = 93.93$ mm,
 $c = 144.4$ mm
(h) $B = 81°54'$,
 $b = 9.947$ m
 $c = 3.609$ m
(i) $A = 53°39'$,
 $a = 2124$ mm,
 $b = 2390$ mm
(j) $A = 45°30'$ or $134°30'$,
 $B = 95°30'$ or $6°30'$,
 $c = 23.7$ m or 2.70 m
(k) $A = 13°51'$,
 $B = 144°2'$,
 $b = 17.2$ m
(l) $A = 86°1'$ or $15°43'$,
 $B = 54°51'$ or $125°9'$,
 $a = 112$ mm or 30.5 mm
(m) $A = 44°46'$,
 $C = 49°57'$,
 $a = 10.69$ m
(n) $B = 93°49'$,
 $C = 36°52'$,
 $b = 30.3$ m
(o) $B = 48°31'$,
 $C = 26°25'$,
 $c = 4.247$ m
2) (a) $c = 10.2$ m,
 $A = 50°11'$,
 $B = 69°49'$
 (b) $a = 11.8$ m,
 $B = 44°42'$,
 $C = 79°18'$
 (c) $b = 4.99$ m,
 $A = 82°24'$,
 $C = 60°18'$
 (d) $A = 38°12'$,
 $B = 81°38'$,
 $C = 60°10'$
 (e) $A = 24°42'$,
 $B = 44°54'$,
 $C = 110°24'$
 (f) $A = 34°42'$,
 $B = 18°6'$,
 $C = 127°12'$
3) 64.00 mm     4) $37°35'$
5) $40.5°$     6) $41°27'$
7) (a) 14.2 m     (b) $142°39'$

8) 60.2 and 32.9 mm
9) 21.2 A          10) 13.4, 14°52′
11) 18.6 A         12) 52°27′
13) A = 55°44′,   B = 76°12′,
    C = 48°4′
14) 14.5 m

## ANSWERS TO CHAPTER 8

### Exercise 8.1

3) 2          4) 3 ohm    5) 100 hours
8) 0.3 N/mm²             11) 0.03 mm

### Exercise 8.2

1) 5                      2) no mode
3) 3, 5 and 8            4) 121.96 ohms
5) 646.6 hours          6) 18.4272 mm
7) 5          8) 84        9) £61
10) Mean = 15.09 ohms,
    Median = 15.755 ohms
11) 118.9, 117.2, 119.1 ohms
12) 20.063, 20.095, 20.115 mm
13) 5.5 and 11.5
14) 109.095 mm    15) 23.865 kg
16) 9.6486 m      17) 19.966 kN
18) 641.5 hours

### Exercise 8.3

1) $\bar{x} = 15.99$, $\sigma = 0.01414$
2) $\bar{x} = 11.4925$ N/mm²,
   $\sigma = 0.01452$ N/mm²
3) $\bar{x} = 10.81\%$, $\sigma = 1.2782\%$
4) $\bar{x} = 99.93$ W, $\sigma = 0.17$ W
5) $\bar{x} = 12.66$, $\sigma = 2.98$
6) $\bar{x} = 43.05$, $\sigma = 4.965$

## ANSWERS TO CHAPTER 9

### Exercise 9.1

1) 8.8 mm              2) 0.0128 m²
3) (a) 1200 mm²
   (b) 276 mm²  (c) 261 mm²
   (d) 774 mm²  (e) 1050 mm²
4) 2800 mm²           5) 8900 mm²
6) 2.12 m             7) 3060 mm²
8) (a) 1380 mm²       (b) 6500 mm²
9) (a) 3.31 mm²       (b) 19.3 mm²
10) 157 mm       11) 29.9 mm
12) (a) 11 200 mm²     (b) 302 mm²
13) 34.1 mm       14) 2592 mm²
15) 909

16) (a) 9.05 m²    (b) 7.54 m
17) 1910 mm
18) (a) 5856 m²    (b) 792.5 m²
19) (a) 3.80 m²    (b) 4.19 m²
    (c) 0.39 m²
20) 34.6 m²

### Exercise 9.2

1) 0.108 m³          2) 335 mm
3) 0.008 75 m³       4) 39 000 m³
5) 60 m³             6) 477 mm
7) 1.51 m²
8) 128 000 mm³, 11 700 mm²
9) 7.92 m³          10) 19.9 ℓ
11) 75.4 mm
12) 5.33 m³, 20.5 m²
13) 0.437 m³
14) (a) 0.366 m³ (b) 0.583 m
15) (a) 20.6 m³  (b) 33.8 m²
16) 0.004 19 m³, 0.126 m²
17) (a) 89.8 m³  (b) 76.9 m²
18) (a) 92.1 m³  (b) 50.3 m²
19) 348 m³, 119 m²      20) 1.47 ℓ
21) 943 mm², 110 mm  22) 1.80 m
23) 0.0410 m², 0.725 m
24) $1.35 \times 10^6$ ℓ
25) 14 m³, 22.7 m²
26) (a) 36.7 m³   (b) 80 700 kg
27) 3.03 m³, 8.36 m²     28) 16.3 ℓ
29) 93.7 m²              30) 150 mm
31) 10.5 m³, 18.8 m²     32) 232 m²

### Exercise 9.3

1) 752        2) 172        3) 0.8
4) 99         5) 25.5       6) 1090 m²
7) 17.2 m³
8) 1060 m², 9.46 m/s
9) $3 \times 10^6$ m³ approx.  10) 7.14 m³
11) 44.30 m³
12) 0.0247 m³

## ANSWERS TO CHAPTER 10

### Exercise 10.1

1) 2.21 m³, 22.1 m²      2) 132 m³
3) 2.18 kg           4) 18.8 m²
5) 41.0 m³           6) 0.837 m³
7) 865 kg
8) 37.2 m², 3.88 m³, 8540 kg
9) 56 300 mm³  10) 1 248 000 mm³

## ANSWERS TO CHAPTER 11

### Exercise 11.1

1) 30.8 m     2) 3.72 m
3) 5.09 m     4) 27°53′
5) 8.91 m
6) (a) 219 m    (b) 140 m
    (c) 260 m    (d) S32°42′E
7) 14.1 m     8) 50°58′
9) 1233 m     10) 190.8 m
11) 74.9 m, 20°41′
12) 11.8 m     13) 36.1 m
14) BD = DF = AC = CE = 2.66 m,
    GB = FH = 4.41 m,
    BC = CF = 3.59 m,
    GA = EH = 3.70 m
15) 147.6 m     16) 33.3 m
17) 44.4 m, 116°
18) 15.2 m, N66°48′E

### Exercise 11.2

1) $x$   0     47.55    29.39 − 29.39
   $y$   50.00   15.45 − 40.45 − 40.45
   $x$ − 47.55
   $y$   15.45
2) $x$   34.64   20.00 − 34.64 − 20.00
   $y$   20.00 − 34.64 − 20.00   34.64
3) $x$   35.24    5.22 − 18.75
   $y$   12.83 − 37.14   32.48
4) $x$ − 19.53   36.94
   $y$   56.73   47.28
5) $x$   14.28   35.72
   $y$    8.57   21.43

### Exercise 11.3

1) 45.79 mm     2) 19.95 mm
3) 20.90 mm     4) 24.98 mm
5) 10°44′

### Exercise 11.4

1) 1.64 mm
2) 1°31′; 13.04 mm; 9.59 mm
3) 65°46′; 29.71 mm
4) 53.01 mm     5) 31.99 mm
6) 4°24′; 25.51 mm
7) 104.98 mm     8) 5.18 mm

### Exercise 11.5

1) 2408 mm     2) 5369 mm
3) 12.63 m     4) 1287 mm
5) 2971 mm     6) 5740 mm

7) 16.01 m     8) 2215 mm
9) 2.887 mm; 53.44 mm
10) 30.53 mm

### Exercise 11.6

1) 2210 mm²     2) 3170 mm²
3) 2765 mm each side
4) 540 m²     5) 738 m²
6) (a) 7.55 m²    (b) 8.06 m²
7) 13.4 m²     8) 962 mm²
9) (a) 143 m²    (b) 53.7 m²
    (c) 43.6 m²
10) 11.9 m²     11) 28 m²
12) 8900 mm²     13) 2.12 m
14) 3060 mm²
15) (a) 13.8 m²    (b) 65.0 m²
16) (a) 3.31 m²    (b) 19.3 m²
17) 15.7 m

## ANSWERS TO CHAPTER 12

### Exercise 12.1

1) (b) 130 mm    (d) 153 mm
2) (a) 40 mm    (c) 71.6°
    (d) 126 mm    (e) 5060 mm²
    (f) 26 640 mm²
3) 127 mm     4) 45.3 mm
5) 8.07 m, 10.1 m
6) (a) 1524 m    (b) 43°58′
7) 19.7 m²
8) (a) 13.9 m    (b) 231 m²
9) 261 m²
10) (a) 45.2 m    (b) 927 m²
11) 34.3 m     12) 185 m
13) 108 m²     14) 43.0 m²
15) 19°29′

## ANSWERS TO CHAPTER 13

### Exercise 13.1

1) (a) $\dfrac{1}{1}, \dfrac{2}{3}, \dfrac{11}{16}, \dfrac{13}{19}, \dfrac{37}{54}, \dfrac{87}{127}$

   (b) $\dfrac{1}{1}, \dfrac{1}{2}, \dfrac{2}{3}, \dfrac{3}{5}, \dfrac{17}{28}, \dfrac{37}{61}$

   (c) $\dfrac{1}{2}, \dfrac{12}{25}, \dfrac{13}{27}, \dfrac{38}{79}, \dfrac{1343}{2792}$

   (d) $\dfrac{1}{1}, \dfrac{1}{2}, \dfrac{6}{11}, \dfrac{49}{90}, \dfrac{55}{101}, \dfrac{159}{292}, \dfrac{373}{685}$

2) (a) $\dfrac{7}{1}, \dfrac{8}{1}, \dfrac{31}{4}, \dfrac{287}{37}, \dfrac{318}{41}, \dfrac{923}{119}$

(b) $\dfrac{2}{1}, \dfrac{11}{5}, \dfrac{13}{6}, \dfrac{89}{41}$

(c) $\dfrac{2}{1}, \dfrac{3}{1}, \dfrac{5}{2}, \dfrac{88}{35}, \dfrac{269}{107}, \dfrac{626}{249}$

3) (a) $\dfrac{1}{3}, \dfrac{1}{4}, \dfrac{2}{7}, \dfrac{13}{46}, \dfrac{15}{53}, \dfrac{283}{1000}$

(b) $\dfrac{1}{1}, \dfrac{51}{52}, \dfrac{52}{53}, \dfrac{103}{105}, \dfrac{155}{158}$,

$\dfrac{413}{421}, \dfrac{981}{1}$

(c) $\dfrac{1}{1}, \dfrac{2}{1}, \dfrac{7}{4}, \dfrac{30}{17}, \dfrac{127}{72}, \dfrac{157}{89}, \dfrac{441}{250}$

(d) $\dfrac{3}{1}, \dfrac{4}{1}, \dfrac{19}{5}, \dfrac{23}{6}, \dfrac{134}{35}, \dfrac{157}{41}, \dfrac{291}{76}$,

$\dfrac{739}{193}, \dfrac{1030}{269}, \dfrac{3829}{1000}$

4) $\dfrac{3}{1}, \dfrac{22}{7}, \dfrac{509}{162}, \dfrac{531}{109}, \dfrac{1571}{500}$

5) (a) $2\frac{2}{15}$; $19.2°$

(b) $5\frac{13}{33}$; $48°32'44''$

(c) $8\frac{24}{29}$; $79°26'54''$

(d) $11\frac{11}{16}$; $105°11'15''$

6) $\dfrac{64}{72}$; $225.0$ mm

7) $\dfrac{32 \times 32}{28 \times 56}$ (other combinations are possible to give 32/49)

8) 70/65; 0.004 mm too large

9) 35/85; 2.0588 mm

10) Any combination to give $\dfrac{7}{48}$;

0.729 17 mm

## ANSWERS TO CHAPTER 14

### Exercise 14.1

1) $x^2 + 3x + 2$    2) $2x^2 + 11x + 15$
3) $6x^2 + 16x + 8$    4) $x^2 - 6x + 8$
5) $3x^2 - 11x + 10$
6) $6x^2 - 17x + 5$    7) $x^2 + 2x - 3$
8) $x^2 - 2x - 15$    9) $3x^2 + 13x - 30$
10) $12x^2 + 4x - 21$
11) $2p^2 - 7pq + 3q^2$
12) $6v^2 - 5uv - 6u^2$
13) $x^2 - 9$    14) $4x^2 - 9$

15) $x^2 + 2x + 1$    16) $4x^2 + 12x + 9$
17) $x^2 - 2x + 1$    18) $4x^2 - 12x + 9$
19) $4a^2 + 12ab + 9b^2$
20) $x^2 + 2xy + y^2$    21) $a^2 - 2ab + b^2$
22) $9x^2 - 24xy + 16y^2$

### Exercise 14.2

1) $(x+1)(x+3)$    2) $(x+2)(x+4)$
3) $(x-1)(x-2)$    4) $(x+5)(x-3)$
5) $(x+7)(x-1)$    6) $(x+2)(x-7)$
7) $(x+y)(x-3y)$
8) $(2x+3)(x+5)$
9) $(p+1)(3p-2)$
10) $(2x+1)(2x-6)$
11) $(m+2)(3m-14)$
12) $(3x+1)(7x+10)$
13) $(2a+5)(5a-3)$
14) $(2x+5)(3x-7)$
15) $(2p+3q)(3p-q)$
16) $(4x+y)(3x-2y)$
17) $(x+y)^2$    18) $(2x+3)^2$
19) $(p+2q)^2$    20) $(3x+1)^2$
21) $(m-n)^2$    22) $(5x-2)^2$
23) $(x-2)^2$    24) $(m+n)(m-n)$
25) $(2x+y)(2x-y)$
26) $(3p+2q)(3p-2q)$
27) $(x+1/3)(x-1/3)$
28) $(1+b)(1-b)$
29) $(1/x+1/y)(1/x-1/y)$
30) $(11p+8q)(11p-8q)$

### Exercise 14.3

1) $x^2 - 4x + 3 = 0$
2) $x^2 + 2x - 8 = 0$
3) $x^2 + 3x + 2 = 0$
4) $x^2 - 2.3x + 1.12 = 0$
5) $x^2 - 1.07x - 4.53 = 0$
6) $x^2 + 7.32x + 12.19 = 0$
7) $x^2 - 1.4x = 0$
8) $x^2 + 4.36x = 0$
9) $x^2 - 12.25 = 0$
10) $x^2 - 8x + 16 = 0$

### Exercise 14.4

1) $\pm 6$    2) $\pm 1.25$
3) $\pm 1.333$    4) $-4$ or $-5$
5) 8 or $-9$    6) 2 or $\frac{1}{3}$
7) 3    8) 4 or $-8$
9) $\frac{4}{7}$ or $\frac{3}{2}$    10) $\frac{7}{3}$ or $-\frac{4}{3}$
11) 1.175 or $-0.425$
12) $\frac{5}{6}$ or $\frac{1}{6}$

13) 0.573 or $-2.907$
14) 0.211 or $-1.354$
15) 1 or $-0.2$
16) 3.886 or $-0.386$
17) 0.956 or $-1.256$
18) 2.388 or 0.262
19) 0.44 or $-3.775$
20) 8.385 or $-2.385$
21) $-0.225$ or $-1.775$
22) 11.14 or $-3.14$
23) 1.303 or $-2.303$
24) $\pm53.67$
25) 5.24 or 0.76
26) $-3.064$ or $-0.935$

## Exercise 14.5

1) 149.6        2) 92.4 mm
3) 6.5 m, 9.5 m    4) 40 mm
5) 0.685 or 23.3 m
6) 30 or 72 mm   7) 54.6 mm
8) 50         9) 2.88 m
10) 94.6 × 94.6 mm
11) 2.41 and $-0.41$ s
12) 0.412

## Exercise 14.6

1) $x = 0.333, y = -1.667$;
   $x = 2, y = 5$
2) $x = -1, y = 6$;
   $x = 0.375, y = -0.875$
3) $x = 1.175, y = 2.175$;
   $x = -0.425, y = 0.575$
4) $x = -0.969, y = 0.956$;
   $x = 0.579, y = -1.256$
5) $x = 0.211, y = -1.422$;
   $x = -1.354, y = 1.708$
6) $x = 0.167, y = 0.805$;
   $x = 0.833, y = 2.278$
7) 12 or 14.7 m
8) 2 m, 8 m

# ANSWERS TO CHAPTER 15

## Exercise 15.1

1) $\log_a n = x$    2) $\log_2 8 = 3$
3) $\log_5 0.04 = -2$
4) $\log_{10} 0.001 = -3$
5) $\log_x 1 = 0$    6) $\log_{10} 10 = 1$
7) $\log_a a = 1$    8) $\log_e 7.39 = 2$
9) $\log_{10} 1 = 0$   10) 3

11) 3     12) 4     13) 3
14) 9     15) 64    16) 100
17) 1     18) 2     19) 3
20) $\frac{1}{2}$     21) 1

## Exercise 15.2

1) 2.08         2) 368
3) 0.795        4) 0.000 179
5) 0.267        6) 1.77
7) $-7.38$      8) $-0.322$
9) 4.63         10) 2.35
11) $-2.28$     12) $-36.5$
13) 22.2

## Exercise 15.3

1) (a) 15.8       (b) 1.97
   (c) 3.00       (d) 0.0314
   (e) 0.581      (f) 0.925
2) (a) 853        (b) 39.3
   (c) 2.18
3) 0.003 57      4) 0.0164
5) 0.590         6) 25.7
7) 132           8) 0.741
9) 44.2          10) 0.005 49
11) 0.769

## Exercise 15.4

1) $y = 1.82, x = 0.84$
2) $x = -1.39, -0.5$
3) (a) $T = 631$   (b) $s = 1.12$
4) 4.34 mA per second
5) (a) 0.18 seconds
   (b) 1200 volts per second
6) (a) 89 800 cells
   (b) 1.35 hours
   (c) 27 900 cells/hour
7) (a) $1.10 \times 10^{-8}$ grams
   (b) $10^{-10}$ grams
   (c) 0.578 hours
8) 0.0996

# ANSWERS TO CHAPTER 16

## Exercise 16.1

1) b      2) d      3) e
4) a      5) c      6) b
7) a      8) b      9) b
10) $x^2 + y^2 = 10$

## Exercise 16.2

1) $a = 70$, $b = 50$

2) $k_m = 0.016 + \dfrac{0.023}{\mu}$

3) Gradient = 1500, Intercept = 0

4) $m = 4.5$, $c = 0.5$

5) $k = 0.2$

6) $a = 0.761$, $b = 10.1$

7) $m = 0.040$, $c = 0.20$

8) $m = 0.1$, $c = 1.4$

9) $a = 12.3$, $b = 0.53$

## Exercise 16.3

1) $a = 3$, $n = 2$

2) $a = 2 \times 10^{-6}$, $n = 4$

3) $n = 2$, $R = 10$

4) $n = 4$, for $V = 80$ read $V = 70$

5) $t = 0.3 \text{ m}^{1.5}$

6) $k = 100$, $n = -1.2$

# ANSWERS TO CHAPTER 17

## Exercise 17.1

1) (a) $2 \times 2$      (b) $2 \times 1$
   (c) $3 \times 3$      (d) $2 \times 4$

2) (a) 9    (b) 8      (c) $n^2$

3) (a) $\begin{pmatrix} 1 & 3 \\ 2 & 4 \end{pmatrix}$    (b) $(5 \quad -6)$

   (c) $\begin{pmatrix} a & 2 & x \\ b & 3 & -6 \\ 4 & 5 & 0 \end{pmatrix}$   (d) $\begin{pmatrix} 1 & 6 \\ -2 & 2 \\ -3 & 0 \\ -4 & -1 \end{pmatrix}$

4) (a) $\begin{pmatrix} 0 & 0 \\ 9 & 2 \end{pmatrix}$    (b) $\begin{pmatrix} 4 & 2 \\ -3 & 2 \end{pmatrix}$

   (c) $\begin{pmatrix} \frac{5}{6} & \frac{1}{2} \\ \frac{5}{6} & 1 \end{pmatrix}$

5) $a = -8$, $b = 5$, $c = 1$

6) $\begin{pmatrix} \frac{1}{3} & \frac{1}{20} \\ \frac{1}{30} & \frac{1}{18} \end{pmatrix}$      7) $\begin{pmatrix} 5 & 8 \\ 12 & -2 \end{pmatrix}$

8) $\begin{pmatrix} 0 \\ 5 \end{pmatrix}$

## Exercise 17.2

1) (a) $\begin{pmatrix} 6 & 0 \\ -4 & 2 \end{pmatrix}$    (b) $\begin{pmatrix} -12 & 3 \\ 9 & -6 \end{pmatrix}$

   (c) $\begin{pmatrix} -6 & 3 \\ 5 & -4 \end{pmatrix}$   (d) $\begin{pmatrix} 18 & -3 \\ -13 & 8 \end{pmatrix}$

2) (a) $\begin{pmatrix} 14 & 0 \\ 8 & -2 \end{pmatrix}$   (b) $\begin{pmatrix} 2 & 1 \\ 3 & 1 \end{pmatrix}$

   (c) $\begin{pmatrix} 5 & 11 \\ 10 & 22 \end{pmatrix}$   (d) $\begin{pmatrix} a & b \\ c & d \end{pmatrix}$

   (e) $\begin{pmatrix} ka & kb \\ kc & kd \end{pmatrix}$

3) (a) $\begin{pmatrix} 7 & 10 \\ 15 & 22 \end{pmatrix}$   (b) $\begin{pmatrix} 3 & -5 \\ 5 & 8 \end{pmatrix}$

   (c) $\begin{pmatrix} 8 & 10 \\ 20 & 18 \end{pmatrix}$   (d) $\begin{pmatrix} 18 & 15 \\ 40 & 48 \end{pmatrix}$

   (e) $\begin{pmatrix} 13 & 10 \\ 40 & 53 \end{pmatrix}$

## Exercise 17.3

1) (a) 24      (b) $-6$      (c) 14

2) (a) $x = 1$, $y = 2$
   (b) $x = 4$, $y = 3$
   (c) $x = 0.5$, $y = 0.75$

## Exercise 17.4

1) $\dfrac{1}{3}\begin{pmatrix} 4 & -5 \\ -1 & 2 \end{pmatrix}$   2) $\begin{pmatrix} 3 & -5 \\ -1 & 2 \end{pmatrix}$

3) $\dfrac{1}{4}\begin{pmatrix} 2 & -2 \\ -1 & 3 \end{pmatrix}$   4) No inverse

5) $\dfrac{1}{320}\begin{pmatrix} 4 & -24 \\ -24 & 224 \end{pmatrix}$

6) No inverse      7) $\dfrac{1}{7}\begin{pmatrix} 2 & -3 \\ 1 & 2 \end{pmatrix}$

8) $\dfrac{1}{13}\begin{pmatrix} 5 & 3 \\ -1 & 2 \end{pmatrix}$   9) $\begin{pmatrix} 1 & -1 \\ 0 & 1 \end{pmatrix}$

10) (a) $\dfrac{1}{2}\begin{pmatrix} 2 & 0 \\ -3 & 1 \end{pmatrix}$   (b) $\begin{pmatrix} 2 & -5 \\ -1 & 3 \end{pmatrix}$

   (c) $\dfrac{1}{2}\begin{pmatrix} 19 & -5 \\ -11 & 3 \end{pmatrix}$

   (d) $\begin{pmatrix} 3 & 5 \\ 11 & 19 \end{pmatrix}$   (e) $\dfrac{1}{2}\begin{pmatrix} 19 & -5 \\ -11 & 3 \end{pmatrix}$

   (f) equal

## Exercise 17.5

1) $x = 6$, $y = -5$

2) $x = 1$, $y = 5$

3) $x = 3$, $y = -1$

4) $x = 2, y = -3$

5) $x = \dfrac{16}{11}, y = \dfrac{19}{11}$

6) $x = 2, y = -5$

## ANSWERS TO CHAPTER 18

### Exercise 18.1

1) $-4, 8$    2) $3$       3) $2$

### Exercise 18.2

1) $2x$        2) $7x^6$        3) $12x^2$
4) $30x^4$     5) $1.5t^2$      6) $2\pi R$
7) $\frac{1}{2}x^{-1/2}$    8) $6x^{1/2}$    9) $x^{-1/2}$
10) $2x^{-1/3}$   11) $-2x^{-3}$   12) $-x^{-2}$
13) $-\frac{3}{5}x^{-2}$   14) $-6x^{-4}$   15) $-\frac{1}{2}x^{-3/2}$
16) $-\frac{1}{3}x^{-3/2}$       17) $-\frac{15}{2}x^{-5/2}$
18) $\frac{3}{10}t^{-1/2}$       19) $-0.01h^{-2}$
20) $-35x^{-8}$        21) $8x - 3$
22) $9t^2 - 4t + 5$     23) $4u - 1$
24) $20x^3 - 21x^2 + 6x - 2$
25) $35t^4 - 6t$    26) $\frac{1}{2}x^{-1/2} + \frac{5}{2}x^{3/2}$
27) $-3x^{-3} + 1$   28) $\frac{1}{2}x^{-1/2} - \frac{1}{2}x^{-3/2}$
29) $3x^2 - \frac{3}{2}x^{-3/2}$
30) $1.3t^{0.3} + 0.575t^{-3.3}$
31) $\frac{9}{5}x^2 - \frac{4}{7}x + \frac{1}{2}x^{-1/2}$
32) $-0.01x^{-2}$
33) $4.65x^{0.5} - 1.44x^{-0.4}$
34) $\frac{3}{2}x^2 + 5x^{-2}$   35) $-6 + 14t - 6t^2$
36) $-5, 19$         37) $39.5, 5, 17$
38) $2.5, 2, 1$

### Exercise 18.3

2) $2.135$          3) $7.48$
4) $5\cos x - 3\sin x$
5) $4\cos\theta + 2\sin\theta$

### Exercise 18.4

1) $\dfrac{x^3}{3} + c$          2) $\dfrac{x^9}{9} + c$

3) $\dfrac{2x^{3/2}}{3} + c$       4) $-\dfrac{1}{x} + c$

5) $-\dfrac{1}{3x^3} + c$        6) $2x^{1/2} + c$

7) $\dfrac{3x^5}{5} + c$          8) $\dfrac{5x^9}{9} + e^x + c$

9) $\dfrac{x^3}{3} + \dfrac{x^2}{2} + 3x + c$

10) $\dfrac{x^4}{2} - \dfrac{7x^2}{2} - 4x + c$

11) $\dfrac{x^3}{3} - \dfrac{5x^2}{2} + 2x^{1/2} - \dfrac{2}{x} + c$

12) $-\dfrac{4}{x^2} + \dfrac{2}{x} + \dfrac{2x^{3/2}}{3} + c$

13) $\dfrac{x^3}{3} - \dfrac{3x^2}{2} + 2x + c$

14) $\dfrac{x^3}{3} + 3x^2 + 9x + c$

15) $\dfrac{4x^3}{3} - 14x^2 + 49x + c$

16) $2\sin x - 3\cos x + c$

### Exercise 18.5

1) $y = \dfrac{x^2}{2} + 1$       2) $46.7$

3) $y = 10 + 3x - x^2$, $10$

4) $y = \dfrac{2x^3}{3} + \dfrac{3x^2}{2} + 2x - \dfrac{1}{6}$

5) $\dfrac{t^3}{3} - 3t^2 + 9t - \dfrac{17}{3}$ [

6) $y = x^2 - 2x + 4$
7) $y = e^x - 0.718$
8) $y = \sin\theta$
9) $y = 4.72 - 3\cos t$

### Exercise 18.6

1) $2.33$     2) $8$       3) $8.67$
4) $4$        5) $3.167$   6) $1.89$
7) $0.667$    8) $0.667$   9) $1$
10) $17.4$

### Exercise 18.7

1) $136$      2) $87$      3) $5.167$
4) $0.667$    5) $3.75$    6) $0$
7) $4$        8) $2$       9) $6.39$
10) $5.21$

# INDEX